T0202869

SpringerBriefs in Mathematics

SpringerBriefs in Mathematics showcases expositions in all areas of mathematics and applied mathematics. Manuscripts presenting new results or a single new result in a classical field, new field, or an emerging topic, applications, or bridges between new results and already published works, are encouraged. The series is intended for mathematicians and applied mathematicians.

More information about this series at http://www.springer.com/series/10030

Gabriel Acosta • Ricardo G. Durán

Divergence Operator
and Related Inequalities

Gabriel Acosta
Department of Mathematics and IMAS
University of Buenos Aires and CONICET
Buenos Aires, Argentina

Ricardo G. Durán
Department of Mathematics and IMAS
University of Buenos Aires and CONICET
Buenos Aires, Argentina

ISSN 2191-8198   ISSN 2191-8201 (electronic)
SpringerBriefs in Mathematics
ISBN 978-1-4939-6983-8   ISBN 978-1-4939-6985-2 (eBook)
DOI 10.1007/978-1-4939-6985-2

Library of Congress Control Number: 2017934519

Mathematics Subject Classification (2010): 26D10, 76D07, 42B20, 46E35, 35Q30

Printed on acid-free paper

This Springer imprint is published by Springer Nature
The registered company is Springer Science+Business Media LLC
The registered company address is: 233 Spring Street, New York, NY 10013, U.S.A.

In memory of our dear friend and colleague,
María Amelia Muschietti.

Preface

Given a domain Ω and a function f with vanishing mean in Ω, is it possible to find a zero trace field $\mathbf{u}$ such that

$$\operatorname{div}\mathbf{u} = f$$

and

$$\|\mathbf{u}\|_{W^{1,p}(\Omega)} \leq C\|f\|_{L^p(\Omega)}?$$

If the answer is positive, how irregular can Ω be? And if it is negative, is it possible to enlarge the space of involved functions in which $\mathbf{u}$ is sought in order to obtain solutions?

This problem, that is, the existence of a right inverse of the *divergence operator*, also called div_p in this book, is strongly related to the Korn inequality and to the Stokes equations among many other relevant problems and inequalities. The purpose of this book is to present in a unified and coherent way an overview of many of these central topics.

Our approach follows somehow our own line of research in these matters collecting most of our production related to the field. Notwithstanding that, many proofs have been improved or shortened and the material organized in a coherent fashion that hopefully helps the reader to get smoothly into the subject. Moreover, at each stage, we tried to survey as many contributions as we could covering both classical papers and recent developments. Naturally, length and time restrictions necessarily imposed a limit. We tried to focus on specific aspects and therefore the reader may - and surely will - find omissions that genuinely should deserve to have a place in this book. Our apologies, in advance, for them.

The book is organized as follows: several equivalences of the problem of existence of a right inverse of the divergence are explored in Chapter 1. Chapter 2 is devoted to present the construction of solutions for div_p elaborated by Bogovskii for star-shaped domains and to show how that result can be generalized to more intricate domains, specifically to John domains. Besides geometrical technicalities, the main ideas are the same and the key tool is ultimately, in both cases, the Calderón-Zygmund singular integrals theory. The last sections of that chapter treat the relation of that problem with improved Poincaré inequalities. Chapter 3, in turn, focuses on

Korn's inequalities beginning with several equivalences, for the so-called *second case* of the inequality, in arbitrary domains. Then, Korn's inequalities on John domains are derived by different alternative approaches using, for instance, the results previously obtained for the divergence operator as well as improved Poincaré inequalities. The latter approach turns out to be useful even for obtaining weighted versions of these inequalities on Hölder-α domains and for domains with external cusps. This is extensively treated along Chapter 4 where also families of counterexamples are exhibited to prove that the presented results are sharp. Moreover, in Chapter 4 more irregular cases are treated. In the last part, cusps are allowed to have rough boundaries, besides the singular point at the tip. Weighted Korn's inequalities are also derived for them using elementary decomposition techniques.

Finally, a brief and even informal derivation of some basic equations of *continuum mechanics* and a short proof delimiting what powers of the distance to a set function belong to the Muckenhoupt's A_p class together with some supplementary material is given in three appendices.

We would like to end this preface with just a few more words. We devoted, in a nonsystematic way, more than twelve years to the issues treated in this book. We did it, as it usually happens, without a master plan in mind and following most of the time personal tastes and interests. Looking back we noticed that a very rich net of connections was simply lost in the vast literature dealing with these topics. A net, maybe not ignored by the expert, we wished to expose clearly to a larger audience along the following pages.

Last but not least, we want to thank our coauthors throughout all these years. Without them, we would have never reached this point. When we were finishing this book we received the sad news that our friend Maria Amelia Muschietti passed away. An important part of the contents presented here corresponds to work done in collaboration with her. This book is dedicated to her memory.

Buenos Aires, Argentina   Gabriel Acosta
December, 2016   Ricardo G. Durán

Contents

Notation and Preliminaries

With $\Omega \subset \mathbb{R}^n$ we denote a general domain (i.e., an open and connected set) which is always assumed to be bounded. The diameter of Ω is denoted with $\delta(\Omega)$. The standard Lebesgue and Sobolev spaces in Ω are denoted respectively with $L^p(\Omega)$ and $W^{k,p}(\Omega)$. Accordingly, we define for $1 \le p < \infty$,

$$\|u\|_{L^p(\Omega)} = \left(\int_\Omega |u|^p dx \right)^{1/p}$$

and recursively

$$\|u\|_{W^{k,p}(\Omega)} = \|u\|_{W^{k-1,p}(\Omega)} + \sum_{|\beta|=k} \|D^\beta u\|_{L^p(\Omega)},$$

where $\beta = (\beta_1, \cdots, \beta_n)$ with $\beta_i \in \mathbb{N}_0$, $|\beta| = \sum_{i=1}^n \beta_i$ and $D^\beta := \frac{\partial^{|\beta|}}{\partial x_1^{\beta_1} x_2^{\beta_2} \cdots x_n^{\beta_n}}$. In this context, Sobolev seminorms are given by

$$|u|_{W^{k,p}(\Omega)} = \sum_{|\beta|=k} \|D^\beta u\|_{L^p(\Omega)}.$$

Sobolev norms and seminorms for $p = \infty$ are defined in the same way adopting usual modifications. We also write $H^k(\Omega) := W^{k,2}(\Omega)$ and use $L^p(\Omega)^n$ and $W^{1,p}(\Omega)^n$ (resp. $L^p(\Omega)^{n \times n}$ and $W^{1,p}(\Omega)^{n \times n}$) for the vectorial (resp. tensorial) version with the natural norm adaptation. In order to help the readability, elements belonging to these spaces are written in bold. Accordingly, we write in a compact form $\|\mathbf{Du}\|_{L^p(\Omega)^{n \times n}}$ with $\mathbf{Du}$ the differential matrix of $\mathbf{u}$, using sometimes $\|\nabla u\|_{L^p(\Omega)^n}$ for scalar functions. Within this context, we use u_i for the components of a vectorial variable $\mathbf{u}$. Accordingly, we denote the divergence with $\text{div}\,\mathbf{u} = \sum_{i=1}^n \frac{\partial u_i}{\partial x_i}$ reserving the symbol Div for the divergence of a matrix field $\mathbf{M}(x)$. For each element in $\mathbf{M}(x) \in W^{1,p}(\Omega)^{n \times n}$, $\text{Div}\,\mathbf{M} \in L^p(\Omega)^n$ and is defined as $\text{Div}\,\mathbf{M}_i = \sum_{j=1}^n \frac{\partial M_{ij}}{\partial x_j}$ (i.e., the operator div is applied along each row of $\mathbf{M}(x)$).

A weight is a nonnegative measurable function defined in Ω. For a weight ω and $1 \leq p < \infty$, we denote with $L^p(\Omega, \omega)$ the *weighted* Lebesgue space with norm $\|f\|_{L^p(\Omega,\omega)}^p = \int_\Omega |f|^p \omega$. Moreover, we define for a couple of weights ω_1, ω_2

$$W^{1,p}(\Omega, \omega_1, \omega_2) := \{f \in L^p(\Omega, \omega_1) : \; \nabla f \in L^p(\Omega, \omega_2)^n\}$$

and the corresponding norm

$$\|f\|_{W^{1,p}(\Omega,\omega_1,\omega_2)}^p := \|f\|_{L^p(\Omega,\omega_1)}^p + \sum_{j=1}^{n} \left\| \frac{\partial f}{\partial x_j} \right\|_{L^p(\Omega,\omega_2)}^p,$$

with similar adaptations to the vectorial and tensorial cases. If both weights agree we write $W^{1,p}(\Omega, \omega) := W^{1,p}(\Omega, \omega, \omega)$.

Functions with zero trace are defined as $W_0^{1,p}(\Omega, \omega_1, \omega_2) := \overline{C_0^\infty(\Omega)}$ and if ω is such that $L^p(\Omega, \omega) \subset L^1(\Omega)$,

$$L_0^p(\Omega, \omega) := \left\{ f \in L^p(\Omega, \omega) : \; \int_\Omega f = 0 \right\}.$$

For a weight $\omega \in L^1(\Omega)$, we write $\omega(\Omega)$ in order to denote $\int_\Omega \omega(x) dx$. However, if $\omega = 1$, we use $|\Omega|$ instead. For a - scalar or vectorial - function $u \in L^1(\Omega, \omega)$ the weighted average $\frac{1}{\omega(\Omega)} \int_\Omega u \omega \, dx$ is denoted with $u_{\Omega,\omega}$ - or just u_ω if Ω is clear from the context. For convenience this notation is straightforwardly extended from weights to arbitrary - even unsigned - functions φ for which $\int_\Omega \varphi \omega = 1$. Accordingly, if Ω is clearly understood from the context, we use $u_{\varphi\omega} = \int_\Omega u \varphi \omega$ that is shortened to $u_\varphi = \int_\Omega u \varphi$ in the case $\omega = 1$.

For a set M, which usually is a part of the boundary of Ω, we write $d_M(x)$ to denote the distance from M to x. In the particular case in which $M = \partial \Omega$, we just write d instead.

Given $1 < p < \infty$, we denote with $W^{-1,p'}(\Omega, \omega^{1-p'})$ the dual space of $W_0^{1,p}(\Omega, \omega)$. Notice that $W^{-1,p}(\Omega, \omega)$ agrees with the dual of $W_0^{1,p'}(\Omega, \omega^{1-p'})$.

For a set $\Gamma \subset \partial \Omega$ with positive $n-1$ Hausdorff measure and such that the trace operator is well defined, $W_\Gamma^{1,p}(\Omega)$ denotes the set of functions in $W^{1,p}(\Omega)$ that vanish on Γ.

For any space $\mathscr{W}$, containing the set of constant functions, we denote the quotient $[\mathscr{W}] := \mathscr{W}/\mathbb{R}$ with the natural norm

$$\|u\|_{[\mathscr{W}]} := \inf_c \|u - c\|_{\mathscr{W}},$$

identifying sometimes $u \in \mathscr{W}$ with the class $[u]$ in order to simplify notation. With $sym(\mathbf{M}) = \frac{1}{2}(\mathbf{M} + \mathbf{M}^t)$ and $skew(\mathbf{M}) = \frac{1}{2}(\mathbf{M} - \mathbf{M}^t)$ we denote the symmetric and skew symmetric part of a matrix $\mathbf{M} \in \mathbb{R}^{n \times n}$ respectively. Since $sym(\mathbf{M})$ appears very frequently it is sometimes shortened to $\boldsymbol{\mu}(\mathbf{M})$. The quotient space $[L^p(\Omega)^{n\times n}]_{Skew}$ is defined as $L^p(\Omega)^{n\times n}/Skew$ with the norm

$$\|\mathbf{M}(x)\|_{[L^p(\Omega)^{n\times n}]_{Skew}} := \inf_{\boldsymbol{sym}(\mathbf{R})=0} \|\mathbf{M}(x) - \mathbf{R}\|_{L^p(\Omega)^{n\times n}}.$$

The scalar product for matrices is defined as $\mathbf{M}:\mathbf{N} = \sum_{i,j=1}^{n} m_{ij}n_{ij} = tr(\mathbf{MN}^t)$, which is associated with the Frobenius norm. Accordingly we have $\|\mathbf{Du}\|_{L^2(\Omega)}^2 = \int_\Omega \mathbf{Du}:\mathbf{Du}$. It is easy to check that $\mathbf{M}:\mathbf{N} = 0$ for any pair of matrices such that $\boldsymbol{sym}(\mathbf{M}) = 0$ and $\boldsymbol{skew}(\mathbf{N}) = 0$. Hence, defining the symmetric part of $\mathbf{Du}$ with $\boldsymbol{\varepsilon}(\mathbf{u})$ we arrive to the following property $\boldsymbol{\varepsilon}(\mathbf{u}):\mathbf{Dv} = \boldsymbol{\varepsilon}(\mathbf{u}):\boldsymbol{\varepsilon}(\mathbf{v})$, used frequently along the following pages.

For every collection of sets $\mathscr{C}$, sometimes we denote with $\cup\mathscr{C}$ the union of all the sets in $\mathscr{C}$, i.e., $\cup\mathscr{C} := \cup_{S\in\mathscr{C}}S$.

Finally, the symbol C denotes a generic constant that may change from line to line and $a \sim b$, sometimes written $a \underset{C}{\sim} b$, means that a and b are comparable, that is $\frac{1}{C}a \leq b \leq Ca$.

Chapter 1
Introduction

The variational analysis of the classical equations of mechanics, both for fluids and elastic solids, is strongly based on some inequalities involving a function and its derivatives.

Here we consider two of these fundamental inequalities. We present them here in order to explain the approach that we are going to use in this work. In other parts of the book we will give more details on the history and different equivalent forms of these inequalities that have been used.

The first inequality, which is a basic tool in the analysis of the Stokes equations, says that, under suitable assumptions on the domain $\Omega \subset \mathbb{R}^n$, there exists a constant C depending only on Ω, such that for any $f \in L_0^2(\Omega)$,

$$\|f\|_{L^2(\Omega)} \leq C\|\nabla f\|_{H^{-1}(\Omega)^n} \tag{1.0.1}$$

Using standard duality arguments it is not difficult to see that this inequality is equivalent to the following result. Given $f \in L_0^2(\Omega)$ there exists a solution $\mathbf{u} \in H_0^1(\Omega)^n$ of the equation

$$\operatorname{div} \mathbf{u} = f \quad \text{in } \Omega \tag{1.0.2}$$

such that

$$\|\mathbf{u}\|_{H^1(\Omega)^n} \leq C\|f\|_{L^2(\Omega)} \tag{1.0.3}$$

where C depends only on Ω.

The second result that we are going to consider is the so-called Korn inequality, which is the basis for the analysis of the classic linear elasticity equations. There are several forms of this inequality but, as we will see, all of them can be deduced from the result given below. In order to state this inequality we need to introduce some notation.

© The Author(s) 2017
G. Acosta, R.G. Durán, *Divergence Operator and Related Inequalities*,
SpringerBriefs in Mathematics, DOI 10.1007/978-1-4939-6985-2_1

For a vector field $\mathbf{v} = (v_1, \cdots, v_n) \in H^1(\Omega)^n$, $\mathbf{Dv}$ denotes the matrix of its first derivatives and $\boldsymbol{\varepsilon}(\mathbf{v})$ its symmetric part (i.e., the strain tensor), namely

$$\varepsilon_{ij}(\mathbf{v}) = \frac{1}{2}\left(\frac{\partial v_i}{\partial x_j} + \frac{\partial v_j}{\partial x_i}\right).$$

Then, the Korn inequality states that, under appropriate assumptions on $\Omega \subset \mathbb{R}^n$, there exists a constant C depending only on Ω such that

$$\|\mathbf{Dv}\|_{L^2(\Omega)^{n\times n}} \leq C\{\|\mathbf{v}\|_{L^2(\Omega)^n} + \|\boldsymbol{\varepsilon}(\mathbf{v})\|_{L^2(\Omega)^{n\times n}}\}. \tag{1.0.4}$$

Our arguments will be based on representation of a function in terms of its first derivatives. These representation formulas can be seen as generalizations to higher dimensions of the elementary one dimensional result, sometimes called Barrow's rule, which is an immediate consequence of the fundamental theorem of calculus, namely, for a function $\varphi \in C^1(\mathbb{R})$,

$$\varphi(y) - \varphi(0) = \int_0^y \varphi'(t)dt \tag{1.0.5}$$

Before going on let us make some important comments. It is reasonable to expect that representation formulas of a function in terms of its gradient can be useful to obtain results like (1.0.1) and consequently of its dual version (1.0.2). However, it is not clear why they are also useful to prove the Korn inequality (1.0.4). The reason is the following identity

$$\frac{\partial^2 v_i}{\partial x_j \partial x_k} = \frac{\partial \varepsilon_{ik}(\mathbf{v})}{\partial x_j} + \frac{\partial \varepsilon_{ij}(\mathbf{v})}{\partial x_k} - \frac{\partial \varepsilon_{jk}(\mathbf{v})}{\partial x_i}. \tag{1.0.6}$$

Using this relation, one can apply the representation formulas to write first derivatives of $\mathbf{v}$ in terms of second derivatives of it, and therefore, in terms of derivatives of $\boldsymbol{\varepsilon}(\mathbf{v})$ and finally in terms of $\boldsymbol{\varepsilon}(\mathbf{v})$ itself (we will see details in Chapter 3). As far as we know, all the proofs of (1.0.4) use in some way (1.0.6) or some variant of it.

Let us go back to the representation formulas. As it is learned in elementary courses, there is an immediate extension to higher dimensions of formula (1.0.5). Indeed, for a test function $\varphi \in C_0^\infty(\mathbb{R}^n)$, applying (1.0.5) to the restriction of φ to a segment we have

$$\varphi(y) - \varphi(0) = \int_0^1 y \cdot \nabla\varphi(ty)dt \tag{1.0.7}$$

Another possibility is to integrate from the given y to ∞. In this way, making a change of variable to write the integral in a finite interval we obtain

$$\varphi(y) = -\int_0^1 \frac{y}{t} \cdot \nabla\varphi\left(\frac{y}{t}\right)\frac{dt}{t} \tag{1.0.8}$$

Using duality, we can see that these representation formulas give two different solutions of div $\mathbf{u} = f$. One of these solutions is related with the so-called Bogovskii formula which will play a fundamental role in our presentation. The other solution is also well known and was introduced by Poincaré. To illustrate let us show how we obtain the Poincaré solution from (1.0.8). Given $f \in C(\mathbb{R}^n)$ with compact support we have, for any test function $\varphi \in C_0^\infty(\mathbb{R}^n)$,

$$\int_{\mathbb{R}^n} f(y)\varphi(y)dy = - \int_{\mathbb{R}^n} \int_0^1 f(y)\frac{y}{t}\cdot\nabla\varphi\left(\frac{y}{t}\right)\frac{dt}{t}dy$$

Then, applying Fubini's theorem and the change of variable $x = y/t$ we obtain

$$\int_{\mathbb{R}^n} f(y)\varphi(y)dy = - \int_{\mathbb{R}^n} \int_0^1 f(tx)x\cdot\nabla\varphi(x)t^{n-1}dtdx$$

Therefore, defining

$$\mathbf{u}(x) = \int_0^1 f(tx)xt^{n-1}dt \tag{1.0.9}$$

we have

$$\int_{\mathbb{R}^n} f(y)\varphi(y)dy = - \int_{\mathbb{R}^n} \mathbf{u}(x)\cdot\nabla\varphi(x)dx$$

which is the weak form of

$$\text{div}\,\mathbf{u} = f$$

and this is indeed the Poincaré solution.

As we mentioned above, an analogous argument using the representation formula (1.0.7) and assuming now that $\int_{\mathbb{R}^n} f = 0$ gives another solution of the divergence related with the Bogovskii formula. We will give more details in Chapter 2.1.

Since we want to work with functions in Lebesgue spaces which are not well defined on lines, representation formulas like (1.0.7) or (1.0.8) are not useful. Indeed, formulas like (1.0.9) do not make sense for $f \in L^2$. However, simple modifications of the formulas given above can be done in order to obtain volume instead of line integrals. Consider, for example, (1.0.7). The idea is to replace the origin by other points z and then take an average over z. Let $B(0,\rho)$ be a ball centered at the origin and with radius ρ. For $z \in B(0,\rho)$ and $\varphi \in C_0^\infty(\mathbb{R}^n)$ we have

$$\varphi(y) - \varphi(z) = - \int_0^1 (z-y)\cdot\nabla\varphi(y+t(z-y))dt \tag{1.0.10}$$

Taking now $\omega \in C_0^\infty(B(0,\rho))$ such that $\int_{B(0,\rho)} \omega = 1$, multiplying (1.0.10) by ω and integrating in z, we obtain

$$\varphi(y) - \overline{\varphi} = - \int_{\mathbb{R}^n} \int_0^1 (z-y)\cdot\nabla\varphi(y+t(z-y))\omega(z)dtdz \tag{1.0.11}$$

where $\varphi_\omega = \int_{B(0,\rho)} \varphi(z)\omega(z)dz$. Interchanging the order of integration and making the change of variables $x = y+t(z-y)$ we obtain

$$\varphi(y) - \varphi_\omega = - \int_0^1 \int_{\mathbb{R}^n} \frac{(x-y)}{t}\cdot\nabla\varphi(x)\,\omega\left(y+\frac{x-y}{t}\right)dx\frac{dt}{t^n}$$

This representation formula will be the basis to construct Bogovskii's solutions of the divergence. We emphasize once again that this formula involves a volume integration of $\nabla\varphi$.

Inequality (1.0.1) has many applications but probably the most classic one is to the analysis of the stationary Stokes equations which model the displacement of a viscous incompressible fluid at low Reynold's number. Therefore, as a motivation, we start by recalling a classic theorem on existence and uniqueness of the Stokes equations.

Given a domain $\Omega \subset \mathbb{R}^n$, the system of equations modeling the motion of an incompressible fluid contained in Ω is given by

$$\begin{cases} -\text{Div}\,\sigma = \mathbf{g} & \text{in } \Omega \\ \text{div}\,\mathbf{u} = 0 & \text{in } \Omega \end{cases} \tag{1.0.12}$$

where σ is the Cauchy stress tensor, $\mathbf{g}$ is a given datum related to the external volume forces acting on the fluid, and the divergence of a tensor field is defined by taking the divergence row by row. More details on the deduction of these equations are given in Appendix A.

For a Newtonian fluid σ is given by

$$\sigma = 2\mu\boldsymbol{\varepsilon}(\mathbf{u}) - p\mathbf{I}$$

where $\mathbf{u}$ and p represent velocity and pressure, respectively, and $\mu > 0$ is a physical parameter. Then, in this case, the first equation in (1.0.12) becomes $-2\mu\text{Div}\,\boldsymbol{\varepsilon}(\mathbf{u}) + \nabla p = \mathbf{g}$, and therefore, the system (1.0.12) can be rewritten as

$$\begin{cases} -2\mu\text{Div}\,\boldsymbol{\varepsilon}(\mathbf{u}) + \nabla p = \mathbf{g} & \text{in } \Omega \\ \text{div}\,\mathbf{u} = 0 & \text{in } \Omega \end{cases} \tag{1.0.13}$$

which is the so-called Stokes system of equations.

Using that $2\text{Div}\,\boldsymbol{\varepsilon}(\mathbf{u}) = \Delta\mathbf{u} + \nabla(\text{div}\,\mathbf{u})$ and assuming a homogeneous Dirichlet boundary condition we obtain

$$\begin{cases} -\mu\Delta\mathbf{u} + \nabla p = \mathbf{g} & \text{in } \Omega \\ \text{div}\,\mathbf{u} = 0 & \text{in } \Omega \\ \mathbf{u} = 0 & \text{on } \partial\Omega \end{cases} \tag{1.0.14}$$

This is the usual way in which Stokes equations are written in mathematical books and papers. However, it is important to mention that one should be careful when dealing with other kind of boundary conditions, for example Neumann type conditions on all the boundary or on part of it. Indeed, in that case one has to use (1.0.13) and not (1.0.14) in order to obtain the correct physical boundary conditions in the integration by parts leading to the weak formulation (an interesting analysis of this problem is given in [74]). We will give the existence and uniqueness theorem for the Stokes equations with homogeneous Dirichlet boundary condition and will comment in a remark how the arguments can be generalized to other boundary conditions.

We denote with $\langle \mathbf{g}, \mathbf{v}\rangle$ the duality product between $\mathbf{g} \in H^{-1}(\Omega)^n$ and $\mathbf{v} \in H_0^1(\Omega)^n$.

Theorem 1.1. *Let $\Omega \subset \mathbb{R}^n$ be an arbitrary bounded domain. There exists a constant C_1 depending only on Ω, such that, for any $f \in L_0^2(\Omega)$,*

$$\|f\|_{L^2(\Omega)} \le C_1 \|\nabla f\|_{H^{-1}(\Omega)^n}, \tag{1.0.15}$$

if and only if, for any $\mathbf{g} \in H^{-1}(\Omega)^n$, there exists a unique solution $(\mathbf{u}, p) \in H_0^1(\Omega)^n \times L_0^2(\Omega)$ of the Stokes equations (1.0.14), and a constant C_2, depending only on Ω and μ, such that

$$\|\mathbf{u}\|_{H_0^1(\Omega)^n} + \|p\|_{L^2(\Omega)} \le C_2 \|\mathbf{g}\|_{H^{-1}(\Omega)^n} \tag{1.0.16}$$

Proof. Assume that (1.0.15) holds. The weak form of the first equation in (1.0.14) is given by

$$\mu \sum_{i,j=1}^{n} \int_\Omega \frac{\partial u_i}{\partial x_j} \frac{\partial v_i}{\partial x_j} + \int_\Omega p \operatorname{div} \mathbf{v} = \langle \mathbf{g}, \mathbf{v} \rangle \quad \forall \mathbf{v} \in H_0^1(\Omega)^n.$$

Introducing the subspace

$$\mathbf{V} = \{\mathbf{v} \in H_0^1(\Omega)^n : \operatorname{div} \mathbf{v} = 0\}$$

the solution $\mathbf{u}$ should belong to $\mathbf{V}$ and satisfy

$$\mu \sum_{i,j=1}^{n} \int_\Omega \frac{\partial u_i}{\partial x_j} \frac{\partial v_i}{\partial x_j} = \langle \mathbf{g}, \mathbf{v} \rangle \quad \forall \mathbf{v} \in \mathbf{V}. \tag{1.0.17}$$

But, the existence of $\mathbf{u} \in \mathbf{V}$ satisfying (1.0.17) and

$$\mu \|\mathbf{u}\|_{H_0^1(\Omega)^n} \le \|\mathbf{g}\|_{H^{-1}(\Omega)^n}. \tag{1.0.18}$$

follows from the Poincaré inequality and the Lax-Milgram theorem.

It remains to show that there exists $p \in L_0^2(\Omega)$ such that

$$\nabla p = \mathbf{g} + \mu \Delta \mathbf{u}. \tag{1.0.19}$$

Here is where we have to use (1.0.15). It follows from this hypothesis that

$$Im\nabla := \{\mathbf{g} \in H^{-1}(\Omega)^n : \mathbf{g} = \nabla f, \text{with } f \in L_0^2(\Omega)\}$$

is a closed subspace of $H^{-1}(\Omega)^n$. Indeed, suppose that we have a sequence $f_m \in L_0^2(\Omega)$ such that $\nabla f_m \to \mathbf{h}$ in $H^{-1}(\Omega)^n$. Then, applying (1.0.15) to $f_m - f_k$, it follows that f_m is a Cauchy sequence in $L_0^2(\Omega)$, and therefore, there exists $f \in L_0^2(\Omega)$ such that $f_m \to f$ in $L_0^2(\Omega)$. Consequently $\mathbf{h} = \nabla f$.

Now, since

$$\langle \mathbf{g} + \mu \Delta \mathbf{u}, \mathbf{v} \rangle = 0 \quad \forall \mathbf{v} \in \mathbf{V}$$

to show the existence of $p \in L_0^2(\Omega)$ satisfying (1.0.19) it is enough to prove that any $\mathbf{h} \in H^{-1}(\Omega)^n$ satisfying

$$\langle \mathbf{h}, \mathbf{v} \rangle = 0 \quad \forall \mathbf{v} \in \mathbf{V} \tag{1.0.20}$$

belongs to $Im\nabla$.

Suppose that this is not the case, then, there exists $\mathbf{h} \in H^{-1}(\Omega)^n$ satisfying (1.0.20) and such that $\mathbf{h} \notin Im\nabla$. Therefore, since $Im\nabla$ is a closed subspace, by the Hahn-Banach theorem, there exists $\mathbf{w} \in H_0^1(\Omega)^n$ such that $\langle \mathbf{h}, \mathbf{w} \rangle = 1$ and $\langle \nabla q, \mathbf{w} \rangle = 0$, for all $q \in L_0^2(\Omega)$. But, from the last equation it follows that $\mathbf{w} \in \mathbf{V}$. So, taking $\mathbf{v} = \mathbf{w}$ in (1.0.20) we obtain a contradiction with $\langle \mathbf{h}, \mathbf{w} \rangle = 1$.

Finally, the a priori estimate (1.0.16) follows from (1.0.19), (1.0.18), and (1.0.15) with $f = p$.

To prove the converse, given $f \in L_0^2(\Omega)$, we consider problem (1.0.14) with $\mathbf{g} = \nabla f$. The unique solution $(\mathbf{u}, p) \in H_0^1(\Omega)^n \times L_0^2(\Omega)$ is given by $\mathbf{u} = 0$ and $p = f$. Then, (1.0.15) follows immediately from (1.0.16).   □

Remark 1.1. As we have mentioned above, to treat other type of boundary conditions we have to work with formulation (1.0.13). Consequently, the argument given in the theorem can be reproduced but (1.0.17) will be replaced by

$$a(\mathbf{u}, \mathbf{v}) := 2\mu \sum_{i,j=1}^n \int_\Omega \boldsymbol{\varepsilon}_{ij}(\mathbf{u})\boldsymbol{\varepsilon}_{ij}(\mathbf{v}) = \langle \mathbf{g}, \mathbf{v} \rangle \quad \forall \mathbf{v} \in \mathbf{V}.$$

with an appropriate definition of the subspace $\mathbf{V}$ depending on the boundary conditions (we omit details because they are analogous to the case of the elasticity equations that we will treat in Chapter 3). Therefore, the proof can be generalized if we have coercivity of the bilinear form a in the corresponding $\mathbf{V}$. This result is the so-called Korn inequality and is not trivial at all. As we will see in Chapter 3 condition (1.0.15) implies the Korn inequality, and therefore, Theorem 1.1 can be generalized to other boundary conditions.

Due to the previous theorem, condition (1.0.15) has been the object of many works. Actually, (1.0.15) has been presented and analyzed in several equivalent forms. To give a short (and surely incomplete) summary of the history of the subject we recall some of the conditions introduced in classic references. This is done in the following two propositions where we also prove the equivalence between each of those conditions with (1.0.15).

The reason why we divide the results into two propositions is that the equivalences given in the first one are valid for arbitrary bounded domains while, for the result proved in the second one, some smoothness requirement on the domains is needed, at least as far as we know.

We will make use of fractional order Sobolev spaces. In [45], Gagliardo proved that, when $\Omega \subset \mathbb{R}^n$ is a Lipschitz domain, the natural restriction to the boundary of continuous functions can be extended to $H^1(\Omega)$ functions. Moreover, he characterizes the subspace of $L^2(\partial\Omega)$ formed by restrictions of functions in $H^1(\Omega)$. This subspace is now known as $H^{1/2}(\partial\Omega)$ which is a Hilbert space with norm given by

$$\|\phi\|_{H^{1/2}(\partial\Omega)}^2 = \|\phi\|_{L^2(\partial\Omega)}^2 + |\phi|_{H^{1/2}(\partial\Omega)}^2$$

where the seminorm on the right-hand side is defined as

$$|\phi|^2_{H^{1/2}(\partial\Omega)} = \int_{\partial\Omega}\int_{\partial\Omega}\frac{|\phi(x)-\phi(y)|^2}{|x-y|^n}d\sigma(x)d\sigma(y)$$

where $d\sigma$ denotes the surface measure on the boundary.

In the following proposition we will consider arbitrary bounded domains. Therefore, we need to generalize the definition of the fractional order Sobolev space introduced above. This can be done in the following way. For any domain Ω we define the quotient space

$$\mathscr{H}^{1/2}(\partial\Omega) = H^1(\Omega)/H_0^1(\Omega).$$

For $u \in H^1(\Omega)$ we denote with $[u]$ its equivalence class. Then, $\mathscr{H}^{1/2}(\partial\Omega)$ is a Hilbert space with norm given by

$$\|\phi\|_{\mathscr{H}^{1/2}(\partial\Omega)} = \inf_{v\in H_0^1(\Omega)} \|u+v\|,$$

where $\phi = [u]$.

In view of the results in [45], when Ω is Lipschitz, $H^{1/2}(\partial\Omega)$ and $\mathscr{H}^{1/2}(\partial\Omega)$ are isomorphic spaces. Therefore, in that case, in the following proposition we can replace $\mathscr{H}^{1/2}(\partial\Omega)$ by $H^{1/2}(\partial\Omega)$. In one of the statements below we will assume that $\phi = [\mathbf{w}] \in \mathscr{H}^{1/2}(\partial\Omega)^n$ is such that $\int_\Omega \mathrm{div}\,\mathbf{w} = 0$, therefore, it is important to remark that, as it is easily seen, this condition is independent of the choice of the class representative $\mathbf{w}$.

We will also need the following well-known result,

Property C (Compactness): For any bounded domain Ω the inclusion $L^2(\Omega) \subset H^{-1}(\Omega)$ is compact. Indeed, this follows by duality from the compactness of the inclusion $H_0^1(\Omega) \subset L^2(\Omega)$. Let us remark that, due to the zero boundary condition, no assumption on the boundary is required.

Proposition 1.1. *For $\Omega \subset \mathbb{R}^n$ an arbitrary bounded domain the following properties are equivalent.*

1. *$\mathrm{Im}\nabla$ is a closed subspace of $H^{-1}(\Omega)^n$.*

2. *There exists C depending only on Ω such that, for any $f \in L_0^2(\Omega)$,*

$$\|f\|_{L^2(\Omega)} \le C\|\nabla f\|_{H^{-1}(\Omega)^n}.$$

3. *For any $f \in L_0^2(\Omega)$, there exist $\mathbf{u} \in H_0^1(\Omega)^n$ and C depending only on Ω such that*

$$\mathrm{div}\,\mathbf{u} = f \ \text{in}\ \Omega$$

and

$$\|\mathbf{u}\|_{H_0^1(\Omega)^n} \le C\|f\|_{L^2(\Omega)}.$$

4. *For any $\boldsymbol{\phi} = [\mathbf{w}] \in \mathscr{H}^{1/2}(\partial\Omega)^n$ such that $\int_\Omega \mathrm{div}\, \mathbf{w} = 0$, there exist $\mathbf{v} \in H^1(\Omega)^n$ and C depending only on Ω such that*

$$\begin{cases} \mathrm{div}\, \mathbf{v} = 0 & \text{in } \Omega \\ \mathbf{v} = \boldsymbol{\phi} & \text{on } \partial\Omega \end{cases} \tag{1.0.21}$$

and

$$\|\mathbf{v}\|_{H^1(\Omega)^n} \le C\|\boldsymbol{\phi}\|_{\mathscr{H}^{1/2}(\partial\Omega)^n}, \tag{1.0.22}$$

where the boundary condition is a notation meaning that $\boldsymbol{\phi} = [\mathbf{v}]$.

5. *There exists C depending only on Ω such that, for any $f \in L^2(\Omega)$,*

$$\|f\|_{L^2(\Omega)} \le C\left\{\|f\|_{H^{-1}(\Omega)} + \|\nabla f\|_{H^{-1}(\Omega)^n}\right\}.$$

Proof. $(1) \Rightarrow (2)$: The linear operator $\nabla : L_0^2(\Omega) \longrightarrow Im\nabla$ is a continuous bijection. Therefore, since $Im\nabla$ is a complete space, (2) follows from the bounded inverse theorem for Banach spaces.

$(2) \Rightarrow (3)$: Using that $\int_\Omega f = 0$ the equation $\mathrm{div}\, \mathbf{u} = f$ can be written in the equivalent form

$$\langle \nabla\psi, \mathbf{u} \rangle = -\int_\Omega f\psi \qquad \forall \psi \in L_0^2(\Omega). \tag{1.0.23}$$

To show the existence of $\mathbf{u} \in H_0^1(\Omega)^n$ satisfying this equation we first define on $Im\nabla$ the functional

$$L(\nabla\psi) = -\int_\Omega f\psi$$

which is continuous. Indeed, using (2) applied to ψ we have

$$|L(\nabla\psi)| \le \|f\|_{L^2(\Omega)}\|\psi\|_{L^2(\Omega)} \le C\|f\|_{L^2(\Omega)}\|\nabla\psi\|_{H^{-1}(\Omega)^n}.$$

Therefore, by the Hahn-Banach theorem, L can be extended to a continuous functional on $H^{-1}(\Omega)^n$ without increasing its norm. But, by the duality between $H_0^1(\Omega)^n$ and $H^{-1}(\Omega)^n$ and the Riesz representation theorem, there exists $\mathbf{u} \in H_0^1(\Omega)^n$ such that

$$L(\mathbf{g}) = \langle \mathbf{g}, \mathbf{u} \rangle \qquad \forall \mathbf{g} \in H^{-1}(\Omega)^n$$

and

$$\|\mathbf{u}\|_{H_0^1(\Omega)^n} \le C\|f\|_{L^2(\Omega)}.$$

In particular, taking $\psi \in L_0^2(\Omega)$ and $\mathbf{g} = \nabla\psi$, we conclude that $\mathbf{u}$ satisfies $(1.0.23)$.

$(3) \Rightarrow (4)$: In view of the definition of the norm in $\mathscr{H}^{1/2}(\partial\Omega)^n$ as an infimum we can assume that

$$\|\mathbf{w}\|_{H^1(\Omega)^n} \le 2\|\boldsymbol{\phi}\|_{\mathscr{H}^{1/2}(\partial\Omega)^n}. \tag{1.0.24}$$

Now, since we are assuming that div $\mathbf{w} \in L_0^2(\Omega)$, it follows from (3) that there exits $\mathbf{u} \in H_0^1(\Omega)^n$ such that

$$\text{div}\,\mathbf{u} = \text{div}\,\mathbf{w}\ \text{in}\ \Omega$$

and

$$\|\mathbf{u}\|_{H_0^1(\Omega)^n} \leq C\|\text{div}\,\mathbf{w}\|_{L^2(\Omega)}. \tag{1.0.25}$$

Then, $\mathbf{v} := \mathbf{w} - \mathbf{u}$ satisfies (1.0.21). Moreover, the estimate (1.0.22) is a consequence of (1.0.24) and (1.0.25).

$(4) \Rightarrow (1)$: We prove first that $(4) \Rightarrow (3)$. Given $f \in L_0^2(\Omega)$ let $\mathbf{r} \in H^1(\Omega)^n$ be a solution of

$$\text{div}\,\mathbf{r} = f \tag{1.0.26}$$

satisfying

$$\|\mathbf{r}\|_{H^1(\Omega)^n} \leq C\|f\|_{L^2(\Omega)}. \tag{1.0.27}$$

To show that such an $\mathbf{r}$ exists we can, for example, extend f by zero to a ball B containing Ω, solve the Poisson problem in B, namely

$$\begin{cases} \Delta\psi = f & \text{in}\ B \\ \psi = 0 & \text{on}\ \partial B \end{cases}$$

and take $\mathbf{r} := \nabla\psi$. Then, the restriction of $\mathbf{r}$ to Ω satisfies (1.0.26) and (1.0.27). Indeed, (1.0.27) follows from well-known a priori estimates for the Poisson problem in a ball.

Now, from (4), we know that there exists $\mathbf{v} \in H^1(\Omega)^n$ satisfying div $\mathbf{v} = 0$ in Ω, $[\mathbf{v}] = [\mathbf{r}]$, and $\|\mathbf{v}\|_{H^1(\Omega)^n} \leq C\|[\mathbf{r}]\|_{\mathscr{H}^{1/2}(\partial\Omega)^n}$.

Then, since $\|[\mathbf{r}]\|_{\mathscr{H}^{1/2}(\partial\Omega)^n} \leq \|\mathbf{r}\|_{H^1(\Omega)^n}$, $\mathbf{u} : \mathbf{r} - \mathbf{v}$ is the desired solution satisfying (3).

Let us now show that $(3) \Rightarrow (2)$. For $f \in L_0^2(\Omega)$, let $\mathbf{u} \in H_0^1(\Omega)^n$ given by (3). Then,

$$\|f\|_{L^2(\Omega)}^2 = \int_\Omega f\,\text{div}\,\mathbf{u} = -\langle\nabla f, \mathbf{u}\rangle \leq \|\nabla f\|_{H^{-1}(\Omega)^n}\|\mathbf{u}\|_{H_0^1(\Omega)^n} \leq C\|\nabla f\|_{H^{-1}(\Omega)^n}\|f\|_{L^2(\Omega)},$$

and therefore, (2) is proved.

$(2) \Rightarrow (5)$: Fix $\psi \in H_0^1(\Omega)$ such that $\int_\Omega \psi = 1$. The, denoting with f_Ω the average of f over Ω we have

$$f = f - f_\Omega + \int_\Omega (f_\Omega - f)\psi + \int_\Omega f\psi,$$

and so, using the Cauchy-Schwarz inequality,

$$\|f\|_{L^2(\Omega)} \leq \left(1 + \|\psi\|_{L^2(\Omega)}|\Omega|^{\frac{1}{2}}\right)\|f - f_\Omega\|_{L^2(\Omega)} + \left\|\int_\Omega f\psi\right\|_{L^2(\Omega)},$$

but the first term on the right-hand side can be bounded using (2), while for the second one we have

$$\left\| \int_\Omega f\psi \right\|_{L^2(\Omega)} \le \|f\|_{H^{-1}(\Omega)} \|\psi\|_{H^1_0(\Omega)} |\Omega|^{\frac{1}{2}},$$

and therefore (5) is proved.

(5) $\Rightarrow$ (2): Suppose that (2) is not true, then, for every $m \in \mathbb{N}$ there exists $f_m \in L^2_0(\Omega)$ such that

$$\|f_m\|_{L^2(\Omega)} \ge m\|\nabla f_m\|_{H^{-1}(\Omega)^n}.$$

Moreover, modifying f_m dividing by its norm, we can assume that $\|f_m\|_{L^2(\Omega)} = 1$.

Then, from the compactness property **C**, we know that there exists a subsequence f_{m_j} and $f \in H^{-1}(\Omega)$ such that $f_{m_j} \to f$ in $H^{-1}(\Omega)$. But, using (5) we have

$$\|f_{m_j} - f_{m_k}\|_{L^2(\Omega)} \le C\left\{ \|f_{m_j} - f_{m_k}\|_{H^{-1}(\Omega)} + \|\nabla(f_{m_j} - f_{m_k})\|_{H^{-1}(\Omega)^n} \right\},$$

and so, since $\|\nabla f_m\|_{H^{-1}(\Omega)^n} \le 1/m$, we conclude that f_{m_j} is a Cauchy sequence in $L^2_0(\Omega)$.

Therefore, $f_{m_j} \to f$ also in $L^2_0(\Omega)$, and consequently, $\nabla f_{m_j} \to \nabla f$ in $H^{-1}(\Omega)^n$. In particular,

$$\|\nabla f\|_{H^{-1}(\Omega)^n} = \lim_{j\to\infty} \|\nabla f_{m_j}\|_{H^{-1}(\Omega)^n} = 0,$$

and then, f is a constant with vanishing integral, i. e., $f = 0$, and we have arrived at the contradiction

$$0 = \|f\|_{L^2(\Omega)} = \lim_{j\to\infty} \|f_{m_j}\|_{L^2(\Omega)} = 1,$$

and consequently, (2) has to be true.

Finally, (1) is a consequence of (2) as it was already proved in Theorem 1.1. $\square$

Remark 1.2. Since

$$\|\nabla f\|_{H^{-1}(\Omega)^n} = \sup_{0\ne\mathbf{v}\in H^1_0(\Omega)^n} \frac{\langle \nabla f, \mathbf{v}\rangle}{\|\mathbf{v}\|_{H^1_0(\Omega)^n}} = \sup_{0\ne\mathbf{v}\in H^1_0(\Omega)^n} \frac{\int_\Omega f\,\mathrm{div}\,\mathbf{v}}{\|\mathbf{v}\|_{H^1_0(\Omega)^n}},$$

condition (2) can be rewritten as

$$\inf_{0\ne f\in L^2_0(\Omega)} \sup_{0\ne\mathbf{v}\in H^1_0(\Omega)^n} \frac{\int_\Omega f\,\mathrm{div}\,\mathbf{v}}{\|f\|_{L^2(\Omega)}\|\mathbf{v}\|_{H^1_0(\Omega)^n}} > 0.$$

This is why (2) is sometimes called *inf-sup* condition, especially in books or papers on numerical methods for the Stokes equations.

There is still another result which is equivalent to the conditions given in the previous proposition. This is the so-called Lion's lemma and, as far as we know, this was the first form of these equivalent conditions to be published.

For the proof we need a density property which holds under a very general assumption on the domain.

Property D (Density): If $f \in H^{-1}(\Omega)$ and $\nabla f \in H^{-1}(\Omega)^n$, there exists a sequence $f_m \in L^2(\Omega)$ such that $f_m \to f$ in $H^{-1}(\Omega)$ and $\nabla f_m \to \nabla f$ in $H^{-1}(\Omega)^n$.

Proposition 1.2. *Given a bounded domain* $\Omega \subset \mathbb{R}^n$ *satisfying property* **D**, *the following condition is equivalent to those given in Proposition 1.1.*

(6) $\quad f \in H^{-1}(\Omega), \nabla f \in H^{-1}(\Omega)^n \;\Rightarrow\; f \in L^2(\Omega).$

Proof. We will prove that (5) and (6) are equivalent.

(5) $\Rightarrow$ (6): Since we are assuming **D**, given $f \in H^{-1}(\Omega)$ with $\nabla f \in H^{-1}(\Omega)^n$, there exists a sequence $f_m \in L^2(\Omega)$ such that $f_m \to f$ in $H^{-1}(\Omega)$ and $\nabla f_m \to \nabla f$ in $H^{-1}(\Omega)^n$.

Applying (5) to $f_m - f_k$ we have

$$\|f_m - f_k\|_{L^2(\Omega)} \leq C \left\{ \|f_m - f_k\|_{H^{-1}(\Omega)} + \|\nabla(f_m - f_k)\|_{H^{-1}(\Omega)^n} \right\},$$

and consequently, f_m is a Cauchy sequence in $L^2(\Omega)$. Then, there exists $\lim f_m$ in $L^2(\Omega)$, but from uniqueness, this limit has to be equal to f, and therefore, $f \in L^2(\Omega)$ as we wanted to prove.

(6) $\Rightarrow$ (5): The set

$$V(\Omega) = \{f \in H^{-1}(\Omega) : \nabla f \in H^{-1}(\Omega)^n\}$$

with the norm given by

$$\|f\|_{V(\Omega)}^2 = \|f\|_{H^{-1}(\Omega)}^2 + \|\nabla f\|_{H^{-1}(\Omega)^n}^2$$

is a Banach space. Moreover, $L^2(\Omega)$ is continuously embedded in $V(\Omega)$. Therefore, (6) means that $L^2(\Omega)$ and $V(\Omega)$ are the same set, and then, (5) is a consequence of the bounded inverse theorem for Banach spaces. $\quad\square$

The conditions given in Propositions 1.1 and 1.2 were introduced in different contexts by several authors, sometimes in connection with the Stokes equations and other times in the study of Sobolev spaces. Let us give some classic references (surely a non complete list). The numbers refer to those in the above propositions.

The oldest publication containing some of these results seems to be [76] from 1958. Indeed, according to Magenes and Stampacchia, assuming that the domain is C^1, Lions proved (6), as well as a generalization to arbitrary order Sobolev spaces, (see the footnote in [76, Page 320]). This is why this result is sometimes called "Lions lemma." Lions proof is based on a transformation of the problem to a half-space and some extension techniques for Sobolev spaces.

Cattabriga [22, Page 312] proved (2) and (3) (and also their generalization to the L^p case, $1 < p < \infty$) for C^2 three dimensional domains. It is interesting to mention that Cattabriga derived (3) (and by duality (2)), from the existence and stability of solution to the Stokes equations (i.e., he uses the "if" part of Theorem 1.1). To treat the Stokes problem he used the explicit Poisson kernel for the Stokes equations in a half-space.

In [71] Ladyzhenskaya proved (4) for two and three dimensional C^2 domains. This is stated in Problem 2.1 in [71, Page 24]. Actually, that statement does not contain explicitly the estimate (1.0.22) but it can be derived from the arguments. Indeed, the author mentioned in [71, Page 26] that the constructed solution of (1.0.21) belongs to $H^1(\Omega)^n$ if $\phi \in H^{1/2}(\partial\Omega)^n$. The construction starts by solving a Neumann problem, and so smoothness of the domain is needed in order to have regularity of the solution of that problem. With similar arguments, and therefore requiring also smoothness of the domain, Babuska and Aziz, in [10, Page 172], proved (3) for two dimensional domains.

Finally, Necas [80, Page 186] proved (5) (and actually a generalization of this result to arbitrary order Sobolev spaces) for n-dimensional Lipschitz domains.

Chapter 2
Divergence Operator

This chapter deals with solutions of the divergence in Sobolev spaces. We will say that a domain $\Omega \subset \mathbb{R}^n$ satisfies div_p if, for any $f \in L_0^p(\Omega)$, there exists $\mathbf{u} \in W_0^{1,p}(\Omega)^n$ such that

$$\mathrm{div}\,\mathbf{u} = f \quad \text{in } \Omega$$

and

$$\|\mathbf{u}\|_{W^{1,p}(\Omega)^n} \leq C\|f\|_{L^p(\Omega)}$$

where the constant C depends only on Ω and p. First we consider the case of domains which are star-shaped with respect to a ball and give the construction introduced by Bogovskiĭ [14]. In his original paper Bogovskiĭ extended the existence of solutions to the case of Lipschitz domains using that this kind of domains can be written as a finite union of star-shaped domains. In the second section, we extend the construction to the class of John domains, this kind of domains includes the Lipschitz ones as well as many domains with fractal boundaries. The construction analyzed here was given in [3]. The proof that we present is a modification of the original one.

Of course, star-shaped domains are a particular case of John domains. The reason why we present first Bogovskiĭ's construction is because it is simpler and allows to present the main ideas with less technical difficulties. On the other hand, the analysis was extended in [47] to generalize the results for right-hand sides in negative order Sobolev spaces. We do not know whether the results in [47] can be extended to John domains.

© The Author(s) 2017
G. Acosta, R.G. Durán, *Divergence Operator and Related Inequalities*,
SpringerBriefs in Mathematics, DOI 10.1007/978-1-4939-6985-2_2

2.1 Solutions of the Divergence on Star-Shaped Domains

Let us begin by recalling the class of star-shaped domains.

Definition 2.1.1 *A bounded open $\Omega \subset \mathbb{R}^n$ is star-shaped with respect to a ball $B \subset \Omega$ if for every $y \in \Omega$ and every $z \in B$ the segment joining y and z is contained in Ω.*

Actually, given an arbitrary domain Ω and $f \in L^1(\Omega)$, Bogovskiĭ's construction gives a solution of $\operatorname{div} \mathbf{u} = f$, but in general, $\mathbf{u}$ will not vanish on the boundary of Ω. However, as we will show, if Ω is star-shaped with respect to a ball, then $\mathbf{u} = 0$ on $\partial\Omega$.

Let $\Omega \subset \mathbb{R}^n$ be a bounded domain with diameter δ. Take $\omega \in C_0^\infty(\Omega)$ such that $\int_\Omega \omega = 1$ and define $G = (G_1, \cdots, G_n)$ as

$$G(x,y) = \int_0^1 \frac{(x-y)}{s} \, \omega\left(y + \frac{x-y}{s}\right) \frac{ds}{s^n} \tag{2.1.1}$$

The following lemma gives a bound for $G(x,y)$ that will be fundamental in our subsequent arguments.

Lemma 2.1. *For $y \in \Omega$ we have*

$$|G(x,y)| \le \|\omega\|_{L^\infty(\Omega)} \frac{\delta^n}{(n-1)|x-y|^{n-1}} \tag{2.1.2}$$

Proof. Since $\omega \in C_0^\infty(\Omega)$, it follows that the integrand in (2.1.1) vanishes for $z = y + (x-y)/s \notin \Omega$. Therefore, since $y \in \Omega$, we can restrict the integral defining $G(x,y)$ to those values of s such that $|z-y| \le \delta$, that is, $|x-y|/\delta \le s$, and so,

$$|G(x,y)| \le \delta \|\omega\|_{L^\infty(\Omega)} \int_{|x-y|/\delta}^1 \frac{ds}{s^n}$$

which immediately gives (2.1.2). $\square$

In the next lemmas and theorem we introduce the explicit right inverse of the divergence.

Lemma 2.2. *For any $\varphi \in C_0^\infty(\Omega)$ we define $\varphi_\omega = \int_\Omega \varphi\,\omega$. Then, for $y \in \Omega$ we have*

$$(\varphi - \varphi_\omega)(y) = -\int_\Omega G(x,y) \cdot \nabla\varphi(x)\,dx$$

Proof. Extending by zero we can think $\varphi \in C_0^\infty(\mathbb{R}^n)$. Repeating the arguments given in the introduction, see (1.0.11), we have, for $y \in \Omega$,

$$(\varphi - \varphi_\omega)(y) = \int_\Omega \int_0^1 (y-z) \cdot \nabla\varphi(y + s(z-y))\,\omega(z)\,ds\,dz$$

and interchanging the order of integration and making the change of variable $x = y + s(z - y)$ we obtain

$$(\varphi - \varphi_\omega)(y) = \int_0^1 \int_\Omega \frac{(y-x)}{s} \cdot \nabla\varphi(x)\, \omega\left(y + \frac{x-y}{s}\right) dx \frac{ds}{s^n}$$

and the proof concludes by observing that we can interchange again the order of integration. Indeed, using the bound given in (2.1.2) for G, it is easy to see that the integral of the absolute value of the integrand is finite. $\square$

By duality, we obtain the following fundamental result.

Theorem 2.1. *Let $\Omega \subset \mathbb{R}^n$ be an arbitrary bounded domain. Given $f \in L^1(\Omega)$ such that $\int_\Omega f = 0$, define*

$$\mathbf{u}(x) = \int_\Omega G(x,y)\, f(y)\, dy \qquad\qquad (2.1.3)$$

then,

$$\operatorname{div} \mathbf{u} = f \qquad in \ \Omega$$

Proof. First, observe that, in view of (2.1.2), $\mathbf{u}$ is well defined and all its components belong to L^1_{loc}. In particular, $\operatorname{div} \mathbf{u}$ is well defined in the sense of distributions.

Now, using Lemma 2.2, for $\varphi \in C_0^\infty(\Omega)$ we have

$$\int_\Omega f(y)\varphi(y)\, dy = \int_\Omega f(y)(\varphi - \varphi_\omega)(y)\, dy = -\int_\Omega \int_\Omega f(y) G(x,y) \cdot \nabla\varphi(x)\, dx\, dy$$

and interchanging the order of integration, which can be done using again (2.1.2), we obtain

$$\int_\Omega f(y)\varphi(y)\, dy = -\int_\Omega \mathbf{u}(x) \cdot \nabla\varphi(x)\, dx$$

which concludes the proof. $\square$

Up to this point, we have not imposed any condition on the domain Ω other than boundedness. Assume now that $\Omega \subset \mathbb{R}^n$ is star-shaped with respect to a ball $B \subset \Omega$. The following lemma shows that, if we choose ω supported in B, then the function $\mathbf{u}$ defined in (2.1.3) vanishes on $\partial\Omega$. In principle, this will be true when $f \in L^p(\Omega)$ for some $p > n$ since, in this case, one can see that $\mathbf{u}$ defined in (2.1.3) is continuous. This will be proved in the next proposition. For other values of p, we can proceed by density to show that $\mathbf{u} \in W_0^{1,p}$ once we have proved that $\mathbf{u} \in W^{1,p}$.

In all what follows we extend f by zero outside of Ω, and therefore, we can think that $f \in L^p(\mathbb{R}^n)$ whenever $f \in L^p(\Omega)$, but we will write $f \in L^p(\Omega)$ to emphasize that f vanishes outside Ω. Analogously functions in $C_0^\infty(\Omega)$ will be thought as being in $C_0^\infty(\mathbb{R}^n)$. Moreover, we can make the following important observation.

Remark 2.1. The definition of $\mathbf{u}$ given in (2.1.3) can be extended to every $x \in \mathbb{R}^n$.

Proposition 2.1. *Let $f \in L^p(\Omega)$ for some $p > n$. If Ω is star-shaped with respect to a ball B and $\omega \in C_0^\infty(B)$, then $\mathbf{u}$ defined in (2.1.3) is continuous in $\mathbb{R}^n$ and vanishes outside Ω, in particular, $\mathbf{u}(x) = 0$ for all $x \in \partial\Omega$.*

Proof. First we observe that

$$G(x,y) = 0 \quad \text{whenever} \quad x \notin \Omega, y \in \Omega \tag{2.1.4}$$

Indeed, in that case we have that, $z = y + (x-y)/s \notin B$, for any $s \in [0,1]$. Otherwise, since Ω is star-shaped with respect to B, $x = (1-s)y + sz$ would be in Ω. Therefore, recalling that $\omega \in C_0^\infty(B)$ and the definition of $G(x,y)$ we obtain (2.1.4. Consequently, $\mathbf{u} = 0$ for all $x \notin \Omega$.

Therefore, it is enough to prove continuity of $\mathbf{u}$ in an open bounded set containing $\overline{\Omega}$. Take x and $\bar{x}$ in a neighborhood of $\overline{\Omega}$. We have

$$G(x,y) - G(\bar{x},y) = \int_0^1 \left\{ \frac{(x-y)}{s} \omega\left(y + \frac{x-y}{s}\right) - \frac{(\bar{x}-y)}{s} \omega\left(y + \frac{\bar{x}-y}{s}\right) \right\} \frac{ds}{s^n}$$

Now, for y and z varying in a bounded domain, the function $z\omega(y+z)$ is Hölder α, for any $0 < \alpha < 1$, as a function of z, uniformly in y. Therefore, assuming, for example, $|x-y| \le |\bar{x}-y|$, and using that the integrand in the definition of $G(x,y)$ vanishes if $s < |x-y|/\delta$ we obtain

$$|G(x,y) - G(\bar{x},y)| \le C|x-\bar{x}|^\alpha \int_{|x-y|/\delta}^1 \frac{ds}{s^{n+\alpha}} \le C \frac{|x-\bar{x}|^\alpha}{|x-y|^{n-1+\alpha}}$$

with C depending only on δ, n, ω, and α. Then,

$$|\mathbf{u}(x) - \mathbf{u}(\bar{x})| \le C|x-\bar{x}|^\alpha \int_\Omega \frac{|f(y)|}{|x-y|^{n-1+\alpha}} \, dy$$

and the proof concludes by observing that, since $p > n$, we can choose $\alpha > 0$ such that $(n-1+\alpha)p' < n$, and using the Hölder inequality. $\square$

We want to prove that $\mathbf{u} \in W^{1,p}(\Omega)^n$. It is not difficult to prove that $\mathbf{u} \in L^p(\Omega)^n$. Indeed, using the bound (2.1.2) we have

$$|\mathbf{u}(x)| \le C \int_\Omega \frac{|f(y)|}{|x-y|^{n-1}} \, dy,$$

and therefore, the Young inequality implies that $\mathbf{u} \in L^p(\Omega)^n$ and that

$$\|\mathbf{u}\|_{L^p(\Omega)^n} \le C\|f\|_{L^p(\Omega)} \tag{2.1.5}$$

with C depending only on n, δ, and ω.

The difficult part is to show that, for $1 < p < \infty$, $\frac{\partial u_i}{\partial x_j} \in L^p(\Omega)$ whenever $f \in L^p(\Omega)$, and this is our next goal.

A fundamental tool for our arguments is the Calderón-Zygmund singular integral operators theory [19, 20]. Also we will make use of the boundedness of the Hardy-Littlewood maximal operator. For the sake of completeness, we state in the next theorems the results on these subjects that we will use in this section and in the next one. With Σ we denote the unit sphere and with $d\sigma$ the corresponding surface measure.

Theorem 2.2. *Let $K(y,z)$ a function defined for $y \in \mathbb{R}^n$ and $z \in \mathbb{R}^n$, $z \neq 0$ satisfying*

1. $K(y, \lambda z) = \lambda^{-n} K(y, z)$ $\forall \lambda > 0$, $y \in \mathbb{R}^n$, $0 \neq z \in \mathbb{R}^n$
2. $\int_{\Sigma} K(y, \sigma) d\sigma = 0$ $\forall y \in \mathbb{R}^n$.
3. $|K(y,z)| \leq \frac{C_1}{|z|^n}$ $\forall y \in \mathbb{R}^n$ *with* C_1 *independent of* y

   Then, for any $1 < p < \infty$,

$$Tg(y) = \lim_{\varepsilon \to 0} T_\varepsilon g(y)$$

with

$$T_\varepsilon g(y) = \int_{|x-y|>\varepsilon} K(y, x-y) g(x) dx$$

defines a bounded operator in L^p and the convergence holds in the L^p norm. Moreover, there exists a constant C_2, depending on p, n, and C_1 such that, if

$$\tilde{T}g(y) = \sup_{\varepsilon > 0} |T_\varepsilon g(y)|,$$

then

$$\left\| \tilde{T}g \right\|_{L^p} \leq C \|g\|_{L^p}$$

Proof. See [20, Theorem 2]. □

For $g \in L^1_{loc}(\mathbb{R}^n)$ the Hardy-Littlewood maximal operator is defined by

$$Mg(x) = \sup_{r>0} \frac{1}{|B(x,r)|} \int_{B(x,r)} |g(y)| dy.$$

Theorem 2.3. *For any $1 < p < \infty$ there exists a constant C depending only on p and n such that*

$$\|Mg\|_{L^p} \leq C \|g\|_{L^p}$$

Proof. See, for example, [34]. □

In the next lemma we give an expression for $\frac{\partial u_i}{\partial x_j}$ in terms of f. In order to do that we introduce a singular integral operator. It is convenient to introduce $\chi_\Omega(y)$, the characteristic function of Ω, in order to have a kernel which vanishes for y outside Ω. Of course, for f vanishing outside Ω this will not make any change, but to prove the bounds that we will need for the kernel it is important to have $\chi_\Omega(y)$ in its definition. We define

$$T_{ij}g(y) = \lim_{\varepsilon \to 0} \int_{|y-x|>\varepsilon} \chi_\Omega(y) \frac{\partial G_i}{\partial x_j}(x,y) g(x) \, dx.$$

and its adjoint operator T^*_{ij}. The existence of this limit in L^p as well as the continuity of T_{ij}, for $1 < p < \infty$, will be proved below using Theorem 2.2. In particular, it will follow that

$$T^*_{ij} f(x) = \lim_{\varepsilon \to 0} \int_{|y-x|>\varepsilon} \chi_\Omega(y) \frac{\partial G_i}{\partial x_j}(x,y) f(y) \, dy$$

Lemma 2.3. *For $1 \le i, j \le n$ we have*

$$\frac{\partial u_i}{\partial x_j} = T_{ij}^* f + \omega_{ij} f \qquad in \ \Omega \qquad (2.1.6)$$

where

$$\omega_{ij}(y) = \int_{\mathbb{R}^n} \frac{z_i z_j}{|z|^2} \omega(y + z) \, dz \qquad (2.1.7)$$

Proof. From the definition of G_i and using again (2.1.2) to interchange the order of integration we have, for any $\varphi \in C_0^\infty(\Omega)$,

$$-\int_\Omega u_i(x) \frac{\partial \varphi}{\partial x_j}(x) \, dx = -\int_\Omega \int_\Omega G_i(x, y) f(y) \frac{\partial \varphi}{\partial x_j}(x) \, dx \, dy. \qquad (2.1.8)$$

For any $y \in \Omega$,

$$-\int_\Omega G_i(x, y) \frac{\partial \varphi}{\partial x_j}(x) \, dx = -\lim_{\varepsilon \to 0} \int_{|y-x|>\varepsilon} G_i(x, y) \frac{\partial \varphi}{\partial x_j}(x) \, dx$$

$$= \lim_{\varepsilon \to 0} \left\{ \int_{|y-x|>\varepsilon} \frac{\partial G_i}{\partial x_j}(x, y) \varphi(x) \, dx \right.$$

$$\left. - \int_{|y-\zeta|=\varepsilon} G_i(\zeta, y) \varphi(\zeta) \frac{(y_j - \zeta_j)}{|y - \zeta|} \, d\zeta \right\} \qquad (2.1.9)$$

Now, we can decompose the second term on the right-hand side in the following way

$$\int_{|y-\zeta|=\varepsilon} G_i(\zeta, y) \phi(\zeta) \frac{(y_j - \zeta_j)}{|y - \zeta|} \, d\zeta = \varphi(y) \int_{|y-\zeta|=\varepsilon} G_i(\zeta, y) \frac{(y_j - \zeta_j)}{|y - \zeta|} \, d\zeta$$

$$+ \int_{|y-\zeta|=\varepsilon} G_i(\zeta, y)(\varphi(\zeta) - \varphi(y)) \frac{(y_j - \zeta_j)}{|y - \zeta|} \, d\zeta := I_\varepsilon + II_\varepsilon$$

and it is easy to see that $II_\varepsilon \to 0$. Indeed, using the bound given in (2.1.2) for G_i and the fact that φ has bounded derivatives we obtain that there exists a constant C depending only on δ, n and $\|\varphi\|_{W^{1,\infty}(\Omega)}$ such that

$$|II_\varepsilon| \le C\varepsilon$$

On the other hand, we have

$$-\lim_{\varepsilon \to 0} I_\varepsilon = -\lim_{\varepsilon \to 0} \varphi(y) \int_{|y-\zeta|=\varepsilon} \int_0^1 \frac{(\zeta_i - y_i)}{s} \omega \left(y + \frac{\zeta - y}{s} \right) \frac{(y_j - \zeta_j)}{|y - \zeta|} \frac{ds}{s^n} \, d\zeta$$

Then, making the change of variables $r = \varepsilon/s$ and $\sigma = (\zeta - y)/\varepsilon$ and denoting with Σ the unit sphere we obtain

$$- \lim_{\varepsilon \to 0} I_\varepsilon = \varphi(y) \lim_{\varepsilon \to 0} \int_{|y-\zeta|=\varepsilon} \int_\varepsilon^\infty (\zeta_i - y_i) \frac{(\zeta_j - y_j)}{|\zeta - y|} \omega\left(y + r \frac{\zeta - y}{\varepsilon}\right) \frac{r^{n-1}}{\varepsilon^n} \, dr \, d\zeta$$

$$= \varphi(y) \lim_{\varepsilon \to 0} \int_\Sigma \int_\varepsilon^\infty \sigma_i \sigma_j \omega(y + r\sigma) r^{n-1} \, dr \, d\sigma$$

$$= \varphi(y) \lim_{\varepsilon \to 0} \int_\Sigma \int_\varepsilon^\infty \frac{\sigma_i \sigma_j}{|\sigma|^2} \omega(y + r\sigma) r^{n-1} \, dr \, d\sigma$$

$$= \varphi(y) \lim_{\varepsilon \to 0} \int_{|z|>\varepsilon} \frac{z_i z_j}{|z|^2} \omega(y + z) \, dz = \varphi(y) \omega_{ij}(y)$$

and therefore, replacing in (2.1.9) we obtain that, for $y \in \Omega$,

$$- \int_\Omega G_i(x,y) \frac{\partial \varphi}{\partial x_j}(x) \, dx = T_{ij} \varphi(y) + \omega_{ij}(y) \varphi(y), \qquad (2.1.10)$$

Then, (2.1.6) follows from (2.1.10) and (2.1.8). $\square$

Remark 2.2. The previous lemma provides a different way of proving that **u** is a solution of the divergence. Moreover, we can consider $f \in L^1(\Omega)$, not necessarily with vanishing integral, and we have

$$\operatorname{div} \mathbf{u} = f - \left(\int_\Omega f\right) \omega \quad \text{in } \Omega \qquad (2.1.11)$$

Indeed, using the expressions for the derivatives given in Lemma 2.3 and observing that $\sum_{i=1}^n \omega_{ii} = 1$ we have that

$$\operatorname{div} \mathbf{u} = f + \sum_{i=1}^n T_{ii}^* f \quad \text{in } \Omega$$

and so, we have to check that

$$\sum_{i=1}^n T_{ii}^* f = -\left(\int_\Omega f\right) \omega \quad \text{in } \Omega$$

But, we have

$$\sum_{i=1}^n T_{ii}^* f(x) = \lim_{\varepsilon \to 0} \sum_{i=1}^n \int_{|y-x|>\varepsilon} \chi_\Omega(y) \frac{\partial G_i}{\partial x_i}(x,y) f(y) \, dy \qquad (2.1.12)$$

and introducing $\eta_i(y,z) := z_i \omega(y+z)$ we obtain from (2.1.1) that

$$\frac{\partial G_i}{\partial x_i}(x,y) = \int_0^1 \frac{\partial \eta_i}{\partial z_i}\left(y, \frac{x-y}{s}\right) \frac{ds}{s^{n+1}} \qquad (2.1.13)$$

but,

$$\frac{\partial \eta_i}{\partial z_i}(y,z) = \omega(y+z) + z_i \frac{\partial \omega}{\partial z_i}(y+z)$$

and so, making the change of variable $r = 1/s$ in (2.1.13) we obtain

$$\sum_{i=1}^{n} \frac{\partial G_i}{\partial x_i}(x,y) = \sum_{i=1}^{n} \int_1^\infty \frac{\partial \eta_i}{\partial z_i}(y, r(x-y)) \, dr = \int_1^\infty \frac{d}{dr}[r^n \omega(y + r(x-y)] \, dr = -\omega(x)$$

which together with (2.1.12) gives (2.1.11).

Next, we will use the expression given in Lemma 2.3 to prove that $\frac{\partial u_i}{\partial x_j} \in L^p(\Omega)$. The kernel $\chi_\Omega(y) \frac{\partial G_i}{\partial x_j}(x,y)$, and so the operator T_{ij}, can be decomposed in two parts as follows:

$$\chi_\Omega(y) \frac{\partial G_i}{\partial x_j}(x,y) = \int_0^\infty \chi_\Omega(y) \frac{\partial \eta_i}{\partial z_j}\left(y, \frac{x-y}{s}\right) \frac{ds}{s^{n+1}} - \int_1^\infty \chi_\Omega(y) \frac{\partial \eta_i}{\partial z_j}\left(y, \frac{x-y}{s}\right) \frac{ds}{s^{n+1}}$$

$$:= K_1(y, x-y) + K_2(y, x-y)$$

and

$$T_{ij} = T_1 + T_2 \qquad\qquad (2.1.14)$$

with

$$T_\ell g(y) = \lim_{\varepsilon \to 0} \int_{|y-x|>\varepsilon} K_\ell(y, x-y) g(x) \, dx \qquad \text{for} \qquad \ell = 1,2$$

First, we will show that the second part T_2 defines a bounded operator in $L^p(\Omega)$ for $1 \leq p \leq \infty$.

Lemma 2.4. *We have*

$$\|T_2 g\|_{L^p(\Omega)} \leq \frac{(1+\delta)}{n} \|\omega\|_{W^{1,\infty}(\mathbb{R}^n)} |\Omega| \, \|g\|_{L^p(\Omega)} \qquad (2.1.15)$$

Proof. From the definition of η_i we can see that

$$\left| \frac{\partial \eta_i}{\partial z_j}\left(y, \frac{z}{s}\right) \right| \leq \left(1 + \frac{|z|}{s}\right) \|\omega\|_{W^{1,\infty}(\mathbb{R}^n)} \qquad (2.1.16)$$

Now, since supp $\omega \subset B \subset \Omega$ it follows that $\chi_\Omega(y) \frac{\partial \eta_i}{\partial z_j}(y, z/s)$ vanishes for $|z|/s > \delta$. In particular, the integral defining K_2 can be restricted to those values of s such that $s \geq |z|/\delta$, and so, from (2.1.16) we obtain

$$\left| \chi_\Omega(y) \frac{\partial \eta_i}{\partial z_j}\left(y, \frac{z}{s}\right) \right| \leq (1+\delta) \|\omega\|_{W^{1,\infty}(\mathbb{R}^n)}$$

Therefore,

$$|K_2(y,z)| \leq (1+\delta) \|\omega\|_{W^{1,\infty}(\mathbb{R}^n)} \int_1^\infty \frac{ds}{s^{n+1}} = \frac{1+\delta}{n} \|\omega\|_{W^{1,\infty}(\mathbb{R}^n)}$$

and then,

$$|T_2 g(y)| \leq \frac{(1+\delta)}{n} \|\omega\|_{W^{1,\infty}(\mathbb{R}^n)} \int_\Omega |g(x)| \, dx$$

and by the Hölder inequality we obtain (2.1.15). □

In view of the decomposition (2.1.14) it remains to analyze the continuity of T_1. With this goal, we will show, in the next two lemmas, that the kernel $K_1(y,z)$ satisfies the hypotheses of Theorem 2.2.

Lemma 2.5. *We have*

$$|K_1(y,z)| \leq \frac{(1+\delta)}{n} \|\omega\|_{W^{1,\infty}(\mathbb{R}^n)} \frac{\delta^n}{|z|^n} \tag{2.1.17}$$

Proof. By the same arguments used in the proof of Lemma 2.4 we obtain

$$|K_1(y,z)| \leq (1+\delta) \|\omega\|_{W^{1,\infty}(\mathbb{R}^n)} \int_{|z|/\delta}^\infty \frac{ds}{s^{n+1}}$$

which immediately gives (2.1.17). □

Lemma 2.6. $K_1(y,z)$ *is homogeneous of degree* $-n$ *and with vanishing mean value on the unit sphere* Σ, *in the second variable.*

Proof. Given $\lambda > 0$, from the definition of K_1 and making the change of variable $t = s/\lambda$, we have

$$K_1(y, \lambda z) = \int_0^\infty \chi_\Omega(y) \frac{\partial \eta_i}{\partial z_j} \left(y, \frac{\lambda z}{s} \right) \frac{ds}{s^{n+1}} = \lambda^{-n} \int_0^\infty \chi_\Omega(y) \frac{\partial \eta_i}{\partial z_j} \left(y, \frac{z}{t} \right) \frac{dt}{t^{n+1}} = \lambda^{-n} K_1(y,z)$$

On the other hand, making the change of variable $r = 1/s$ in the integral defining K_1 we have

$$K_1(y,z) = \int_0^\infty \chi_\Omega(y) \frac{\partial \eta_i}{\partial z_j} (y, rz) r^{n-1} \, dr$$

and therefore,

$$\int_\Sigma K_1(y,\sigma) \, d\sigma = \int_\Sigma \int_0^\infty \chi_\Omega(y) \frac{\partial \eta_i}{\partial z_j} (y, r\sigma) r^{n-1} \, dr \, d\sigma = \int_{\mathbb{R}^n} \chi_\Omega(y) \frac{\partial \eta_i}{\partial z_j} (y,z) \, dz = 0$$

because $\eta_i(y,z)$ is a smooth function with compact support in the z variable. □

We can now state and prove the main result of this section.

Theorem 2.4. *Let* Ω *be bounded and star-shaped with respect to a ball* $B \subset \Omega$. *If* $f \in L^p(\Omega)$, $1 < p < \infty$, *and* $\int_\Omega f = 0$, *then, the function* **u** *defined in* (2.1.3) *is in* $W_0^{1,p}(\Omega)^n$ *and satisfies*

$$\text{div } \mathbf{u} = f \quad \text{in } \Omega. \tag{2.1.18}$$

Moreover, there exists a constant C depending only on Ω and p, such that

$$\|\mathbf{u}\|_{W_0^{1,p}(\Omega)^n} \leq C\|f\|_{L^p(\Omega)} \tag{2.1.19}$$

Proof. That $\mathbf{u} \in L^p(\Omega)^n$ and

$$\|\mathbf{u}\|_{L^p(\Omega)^n} \leq C\|f\|_{L^p(\Omega)}$$

follows from the definition of $\mathbf{u}$ using (2.1.2) and the Young inequality.

Now we show that, for $1 \leq i, j \leq n$, there exists a constant C depending only on p, δ, n, and ω such that

$$\left\|\frac{\partial u_i}{\partial x_j}\right\|_{L^p(\Omega)} \leq C\|f\|_{L^p(\Omega)}$$

To do that we use the expression for the derivatives given in Lemma 2.3. From (2.1.7) it follows immediately that ω_{ij} is a bounded function. Indeed, $\|\omega_{ij}\|_{L^\infty} \leq \|\omega\|_{L^1}$.

Then, it remains to prove that the operator T_{ij}^* is bounded in L^p, for $1 < p < \infty$. In view of Lemmas 2.5 and 2.6, it follows from Theorem 2.2 that the limit defining T_1 exists in the L^p norm and defines an operator which is continuous in L^p for $1 < p < \infty$. Then, the boundedness of T_{ij} in L^p, for $1 < p < \infty$, follows from the decomposition $T_{ij} = T_1 + T_2$ recalling that, as we proved in Lemma 2.1.15, T_2 is continuous in L^p. By a standard duality argument it follows that T_{ij}^* is also bounded for $1 < p < \infty$. Therefore, we have proved that $\mathbf{u} \in W^{1,p}(\Omega)$ and satisfies

$$\|\mathbf{u}\|_{W^{1,p}(\Omega)^n} \leq C\|f\|_{L^p(\Omega)} \tag{2.1.20}$$

Now, for $p > n$, it follows from Proposition 2.1 that $\mathbf{u}$ is continuous and vanishes on $\partial\Omega$. But, in [85], it is proved that for an arbitrary open set Ω, if a function is continuous and vanishes on $\partial\Omega$ and belongs to $W^{1,p}(\Omega)$, then it belongs to $W_0^{1,p}(\Omega)$.

On the other hand, for $1 < p \leq n$, take a sequence $f_m \in L^\infty(\Omega)$ such that $f_m \to f$ in $L^p(\Omega)$ and let

$$\mathbf{u}_m(x) = \int_\Omega G(x,y) f_m(y) dy.$$

Then, from (2.1.20) applied to $f - f_m$ it follows that $\mathbf{u}_m \to \mathbf{u}$ in $W^{1,p}(\Omega)^n$. But we already know that $\mathbf{u}_m \in W_0^{1,p}(\Omega)^n$, and therefore, $\mathbf{u} \in W_0^{1,p}(\Omega)^n$ and the theorem is proved. $\square$

Remark 2.3. For a Lipschitz domain Ω, the existence of $\mathbf{u}$ satisfying (2.1.18) and (2.1.19) can be proved using the previous theorem and the fact that Ω can be written as a finite union of domains which are star-shaped with respect to a ball. We omit details (which can be found in [14], see also the decomposition technique described in Section 4.5) because in the next section we will generalize Bogovskiǐ's construction to a class of domains which contains the Lipschitz ones.

2.2 Solutions of the Divergence on John Domains

In view of the results of the previous section, an interesting problem is to find weaker sufficient conditions on a bounded domain Ω for the existence of $\mathbf{u}$ satisfying (2.1.18) and (2.1.19).

It is known that the domain cannot be arbitrary, indeed, several counterexamples of domains which do not satisfy this result have been published (see Section 4.4). From these counterexamples it follows that we have to consider a class of domains which excludes domains with external cusps. On the other hand, the Lipschitz condition is not necessary. In fact, it is known that if the result holds for two domains then it also holds for the union of them (see, for example, the argument given in [14]), and consequently, domains having internal cusps are allowed although they are not Lipschitz.

Consequently, it seems that a natural class of domains to be considered for our problem is that of the John domains (see definition below). For instance, it is known that a two dimensional domain with a piecewise smooth boundary is a John domain if and only if it does not have external cusps. These domains were first considered by F. John in his work on elasticity [61] and were named after him by Martio and Sarvas [79]. Further, John domains were used in the study of several problems in Analysis. For example, they were used by G. David and S. Semmes [30] in the analysis of quasiminimal surfaces of codimension one and by S. Buckley and P. Koskela [16] for the study of several inequalities. On the other hand, John domains are closely related with the extension domains of P. Jones [62]. Indeed the (ε, ∞) domains, also called uniform domains, are John domains (but the converse is not true: a John domain can have an internal cusp while a uniform domain cannot).

As we will show in this section, the approach used to construct solutions of the divergence on star-shaped domains can be generalized to John domains. This generalization has been done in [3]. The key idea is to replace the segments used for the integration in (1.0.10) by appropriate curves.

There are several equivalent definitions of John domains. A usual one is the following. We will denote with $d(x)$ the distance from x to $\partial\Omega$.

Definition 2.2.1 *A bounded open $\Omega \subset \mathbb{R}^n$ is a John domain if there exist a positive constant c_1 and $x_0 \in \Omega$ such that, for every $y \in \Omega$ there exists a rectifiable curve $\mathscr{C}_{x_0,y} \subset \Omega$ joining y and x_0 with the following property:*

If $\ell(y)$ denotes the length of $\mathscr{C}_{x_0,y}$ and $\rho : [0, \ell(y)] \to \Omega$ is its parametrization by arc-length such that $\rho(0) = y$, $\rho(\ell(y)) = x_0$, then,

$$d(\rho(t)) \geq c_1 t \quad \forall t \in [0, \ell(y)] \tag{2.2.1}$$

The property given in this definition means that one can reach each point $y \in \Omega$ by a curve $\mathscr{C}_{x_0,y}$ such that any point $x \in \mathscr{C}_{x_0,y}$ is at a distance from the boundary of Ω greater than a fixed proportion of the length of the curve between y and x. This is why this property is sometimes called the "Twisted cone condition."

In some papers condition (2.2.1) is replaced by the following two conditions. There exist two positive constants c_2 and c_3 such that

$$\ell(y) \leq c_2 \tag{2.2.2}$$

and

$$d(\rho(t)) \geq \frac{c_3}{\ell(y)} t \quad \forall t \in [0, \ell(y)] \tag{2.2.3}$$

where c_2 and c_3 are positive constants.

It is not difficult to see that both definitions are equivalent. Indeed, it is obvious that (2.2.2) and (2.2.3) imply (2.2.1) with $c_1 = c_3/c_2$. Conversely, taking $t = \ell(y)$ in (2.2.1) we obtain (2.2.2) with $c_2 = d(x_0)/c_1$. To prove (2.2.3), consider first $\ell(y) < d(x_0)/2$. In this case we have $d(\rho(t)) > d(x_0)/2 \geq d(x_0)t/2\ell(y)$ for all $t \in [0, \ell(y)]$. On the other hand, if $\ell(y) \geq d(x_0)/2$, it follows from (2.2.1) that $d(\rho(t)) \geq c_1 d(x_0)t/2\ell(y)$.

If Ω is a John domain, there is an infinite number of possible choices for the curves satisfying the properties required in Definition 2.2.1. To construct our solution of the divergence we will choose a family of curves verifying some extra conditions, in particular, close to y, the curve joining y and x_0 will be a segment. Moreover, we need to have some control on the variability of the curves as functions of y, indeed, measurability will be enough for our purposes. Also, for convenience we re-scale the curves in order to have the parameter in $[0, 1]$.

In the next lemma we state the properties that we will need and prove the existence of a family of curves satisfying them. We will make use of a Whitney decomposition of an open set. We refer the reader to [84] for a proof of the existence of such a decomposition for any open bounded set.

Definition 2.2.2 *Given* $\Omega \subset \mathbb{R}^n$ *an open bounded set, a* **Whitney decomposition** *of* Ω *is a family* W *of closed dyadic cubes with pairwise disjoint interiors and satisfying the following properties:*

1) $\Omega = \cup_{Q \in W} Q$

2) $diam(Q) \leq d(Q, \partial\Omega) \leq 4\,diam(Q) \quad \forall Q \in W$

3) $\frac{1}{4} diam(Q) \leq diam(\tilde{Q}) \leq 4\,diam(Q) \quad \forall Q, \tilde{Q} \in W \quad such\ that\ Q \cap \tilde{Q} \neq \emptyset$

Given $Q \in W$, let x_Q be its center and Q^* the cube with the same center but expanded by a factor $9/8$, namely, $Q^* = \frac{9}{8}(Q - x_Q) + x_Q$. We will make use of the following facts which follow easily from the properties given in Definition 2.2.2.

$$d(Q^*, \partial\Omega) \sim diam(Q^*) \sim d(y) \quad \forall y \in Q^*, \tag{2.2.4}$$

where $A \sim B$ means that there are constants c and C, which may depend on the dimension n but on nothing else, such that $cA \leq B \leq CA$. We will use the notation $\dot{\gamma}(s, y) := \frac{\partial\gamma}{\partial s}(s, y)$.

Lemma 2.7. *Let $\Omega \subset R^n$ be a bounded John domain and x_0, c_1 and $\rho(t)$ be as in Definition 2.2.1. Then, there exist a function $\gamma : [0,1] \times \Omega \to \Omega$ and positive constants c_J and k depending only on c_1, $diam(\Omega)$, $d(x_0)$ and n, such that*

1) $\gamma(0,y) = y$, $\gamma(1,y) = x_0$

2) $d(\gamma(s,y)) \geq c_J s$

3) $|\dot\gamma(s,y)| \leq c_J^{-1}$

4) $\{x \in \Omega : x = \gamma(s,y), 0 \leq s \leq kd(y)\}$ is a segment and $kd(y) \leq 1$

5) $\gamma(s,y)$ and $\dot\gamma(s,y)$ are measurable functions.

Proof. Let W be a Whitney decomposition of Ω and $Q_0 \in W$ be a cube containing x_0. Given $y \in \Omega$, let $Q \in W$ be such that $y \in Q$. We remark that if y belongs to the boundary of some $Q \in W$ then it belongs to more than one cube. We choose any of them arbitrarily (in any case this is not important because the set of those points is of measure zero).

Suppose first that $x_0 \in Q^*$. In this case we can take the curve as a segment, namely, $\gamma(s,y) = sx_0 + (1-s)y$. In fact, in view of (2.2.4), it is easy to see that $\gamma(s,y)$ satisfies 2) and 3) with c_J depending on $d(x_0)$. Also 4) is trivially satisfied for any k such that $kd(y) \leq 1$, we can take, for example, $k = 1/diam(\Omega)$.

Now, if $x_0 \notin Q^*$, let x_Q be the center of Q and take $\rho(t)$ as a parametrization of a curve joining x_Q and x_0 satisfying the conditions given in the definition of John domains. First we reparametrize ρ and define $\mu(s) = \rho(s\ell(x_Q))$. Then, $d(\mu(s)) \geq c_J s$ for $c_J \sim c_1 \ell(x_Q)$. But, since $x_0 \notin Q^*$, we obtain from properties 2) and 3) of Definition 2.2.2 that $\ell(x_Q) \geq |x_0 - x_Q| \geq cd(x_0)$ with c depending only on n. Therefore, 2) holds for μ with $c_J \sim c_1 d(x_0)$. Moreover, $|\dot\mu(s)| \leq d(x_0)/c_1$ and so we can choose c_J small enough such that μ also satisfies 3).

To define $\gamma(s,y)$ we modify $\mu(s)$ in the following way. Let s_1 be the first $s \in [0,1]$ such that $\mu(s) \in \partial Q^*$. Then we define

$$\gamma(s,y) = \begin{cases} (s/s_1)\mu(s_1) + (1-(s/s_1))y & \text{if } s \in [0, s_1] \\ \mu(s) & \text{if } s \in [s_1, 1] \end{cases}$$

see Figure 2.1.

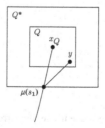

Fig. 2.1 Detail of the ending part of γ

Now, for $s \in [0, s_1]$, $\dot{\gamma}(s, y) = (\mu(s_1) - y)/s_1$. But, since $|\dot{\mu}(s)| \leq d(x_0)/c_1$, $\mu(s_1) \in \partial Q^*$ and $\mu(0) = x_Q$, it is easy to check that $s_1 \geq c c_1 diam(Q^*)/d(x_0)$ with c depending only on n. Therefore, $|\dot{\gamma}(s, y)|$ is bounded by a constant which depends only on n, c_1, and $d(x_0)$. So, we can choose c_J small enough such that $\gamma(s, y)$ satisfies 3) on the interval $[0, s_1]$. On the other hand, for $s \in [0, s_1]$, both $\mu(s)$ and $\gamma(s, y)$ belong to Q^* and so $d(\gamma(s, y)) \sim d(\mu(s))$ and therefore 2) holds on this interval. Since $\gamma(s, y) = \mu(s)$ on $s \in [s_1, 1]$, 2) and 3) hold on all the interval $[0, 1]$.

Using again that $s_1 \geq c c_1 diam(Q^*)/d(x_0)$, 4) follows from (2.2.4).

Finally, observe that 5) holds because $\gamma(s, y)$ and $\dot{\gamma}(s, y)$ are continuous for y in the interior of each $Q \in W$ and so they are continuous up to a set of measure zero. $\quad\square$

Our next goal is to introduce the solution of the divergence which generalizes to John domains that given in (2.1.3). To simplify notation we will assume, without loss of generality, that $x_0 = 0$.

Let c_J be the constant appearing in 2) and 3) of Lemma 2.7 and $\omega \in C_0^\infty$ $(B(0, c_J/2))$ be such that $\int_\Omega \omega = 1$. Given a function $\varphi \in C_0^\infty(\Omega)$ we define $\varphi_\omega = \int_\Omega \varphi \omega$. The key point in our construction is to recover $\varphi - \varphi_\omega$ from its gradient. To do this we replace the segments used in the case of star-shaped domains by appropriate curves based on the function γ defined in Lemma 2.7. Observe that, taking $s = 1$ in 2) of Lemma 2.7, we obtain $c_J \leq d(0)$ and so $B(0, c_J/2) \subset \Omega$.

Now, for any $y \in \Omega$ and any $z \in B(0, c_J/2)$ we define, for $s \in [0, 1]$,

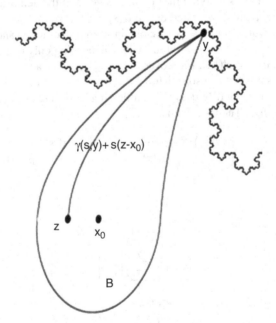

Fig. 2.2 A twisted cone inside of an irregular domain

$$\tilde{\gamma}(s, y, z) = \gamma(s, y) + sz \tag{2.2.5}$$

Then, it follows immediately from 1) of Lemma 2.7 that

$$\tilde{\gamma}(0,y,z) = y \quad \text{and} \quad \tilde{\gamma}(1,y,z) = z \tag{2.2.6}$$

Moreover,

$$\tilde{\gamma}(s,y,z) \in \Omega \qquad \forall s \in [0,1] \tag{2.2.7}$$

Indeed, using 2) of Lemma 2.7, we obtain

$$|\tilde{\gamma}(s,y,z) - \gamma(s,y)| = s|z| < \frac{sc_J}{2} < d(\gamma(s,y))$$

and therefore, (2.2.7) holds (Figure 2.2).

We can now introduce the function $G = (G_1, \cdots, G_n)$ which will be the kernel of the right inverse of the divergence.

For $x \in \mathbb{R}^n$ and $y \in \Omega$ we define

$$G(x,y) = \int_0^1 \left\{ \gamma(s,y) + \frac{x - \gamma(s,y)}{s} \right\} \omega \left(\frac{x - \gamma(s,y)}{s} \right) \frac{ds}{s^n} \tag{2.2.8}$$

Observe that, from 5) of Lemma 2.7, we know that $G(x,y)$ is a measurable function.

Remark 2.4. The integral defining $G(x,y)$ can be restricted to $s \geq c|x-y|$ for c depending only on the constant c_J given in Lemma 2.7. Indeed, if $(x - \gamma(s,y))/s$ is in the support of ω, then

$$|x-y| \leq |x - \gamma(s,y)| + |\gamma(s,y) - \gamma(0,y)| \leq c_J s + \sqrt{n} c_J^{-1} s.$$

An important consequence of this remark is the bound for $G(x,y)$ given in the following lemma.

Lemma 2.8. *There exists a constant* $C = C(n,c_J,\omega)$ *such that*

$$|G(x,y)| \leq \frac{C}{|x-y|^{n-1}} \tag{2.2.9}$$

Proof. In view of Remark 2.4 we have

$$G(x,y) = \int_{c|x-y|}^1 \left\{ \gamma(s,y) + \frac{x - \gamma(s,y)}{s} \right\} \omega \left(\frac{x - \gamma(s,y)}{s} \right) \frac{ds}{s^n}$$

But, for $(x - \gamma(s,y))/s$ in the support of ω, we have

$$\left| \gamma(s,y) + \frac{x - \gamma(s,y)}{s} \right| \leq c_J^{-1} + \frac{c_J}{2}$$

where we have used property 3) of Lemma 2.7. Then, the integrand is bounded by $(c_J^{-1} + c_J/2)\|\omega\|_\infty s^{-n}$ and the estimate (2.2.9) follows by an elementary integration.
$\square$

The next lemma shows how $\varphi - \varphi_\omega$ can be recovered from its gradient by means of the kernel G. As a consequence of this result we obtain our solution of the divergence.

Lemma 2.9. *For $\varphi \in C^1(\Omega) \cap W^{1,1}(\Omega)$ we have that, for any $y \in \Omega$,*

$$(\varphi - \varphi_\omega)(y) = -\int_\Omega G(x,y) \cdot \nabla \varphi(x) dx.$$

Proof. Since $\int_\Omega \omega = 1$ we have, in view of (2.2.6), that for any $y \in \Omega$,

$$(\varphi - \varphi_\omega)(y) = \int_\Omega (\varphi(y) - \varphi(z)) \omega(z) dz = -\int_\Omega \int_0^1 \dot{\tilde{\gamma}}(s,y,z) \cdot \nabla \varphi(\tilde{\gamma}(s,y,z)) \omega(z) ds dz.$$

From (2.2.5) we obtain $\dot{\tilde{\gamma}}(s,y,z) = \dot{\gamma}(s,y) + z$. Then, interchanging the order of integration and making the change of variables $x = \tilde{\gamma}(s,y,z)$, we have $z = (x - \gamma(s,y))/s$ and $dz = dx/s^n$, and hence,

$$(\varphi - \varphi_\omega)(y) = -\int_\Omega \int_0^1 \left\{ \dot{\gamma}(s,y) + \frac{x - \gamma(s,y)}{s} \right\} \omega\left(\frac{x - \gamma(s,y)}{s}\right) \frac{ds}{s^n} \cdot \nabla \varphi(x) dx$$

which in view of the definition (2.2.8) concludes the proof □

Theorem 2.5. *For $f \in L^1(\Omega)$ such that $\int_\Omega f = 0$ define*

$$\mathbf{u}(x) = \int_\Omega G(x,y) f(y) dy \qquad (2.2.10)$$

then

$$\operatorname{div} \mathbf{u} = f.$$

Proof. The proof is exactly as that of Theorem 2.1, using now (2.2.9) to see that $\mathbf{u}$ is well defined and all its components belong to L^1_{loc}, and therefore, $\operatorname{div} \mathbf{u}$ is well defined in the sense of distributions.

Remark 2.5. As in the case of star-shaped domains, the definition of $\mathbf{u}$ given in (2.2.10) can be extended to every $x \in \mathbb{R}^n$.

Proposition 2.2. *Let $f \in L^p(\Omega)$ for some $p > n$. Then, $\mathbf{u}$ defined in (2.2.10) is continuous in $\mathbb{R}^n$ and vanishes outside Ω, in particular, $\mathbf{u}(x) = 0$ for all $x \in \partial\Omega$.*

Proof. First we observe that

$$G(x,y) = 0 \quad \text{whenever } x \notin \Omega, y \in \Omega$$

Indeed, it is enough to see that

$$\omega\left(\frac{x - \gamma(s,y)}{s}\right) = 0 \qquad \text{for } x \notin \Omega \text{, } y \in \Omega \text{ and } s \in [0,1] \qquad (2.2.11)$$

But, in this case we have from property 2) of Lemma 2.7,

$$c_J s \leq d(\gamma(s,y)) \leq |\gamma(s,y) - x|$$

hence,

$$\frac{|\gamma(s,y) - x|}{s} \geq c_J$$

and therefore, (2.2.11) follows immediately since supp $\omega \subset B(0, c_J/2)$. Consequently, $\mathbf{u} = 0$ for all $x \notin \Omega$. Take now x and $\bar{x}$ in a neighborhood of $\overline{\Omega}$. We have

$$G(x,y) - G(\bar{x},y) = \int_0^1 \left\{ \left(\dot{\gamma}(s,y) + \frac{x - \gamma(s,y)}{s} \right) \omega \left(\frac{x - \gamma(s,y)}{s} \right) \right.$$
$$\left. - \left(\dot{\gamma}(s,y) + \frac{\bar{x} - \gamma(s,y)}{s} \right) \omega \left(\frac{\bar{x} - \gamma(s,y)}{s} \right) \right\} \frac{ds}{s^n}$$

and then, since $|\dot{\gamma}(s,y)| \leq c_J^{-1}$,

$$|G(x,y) - G(\bar{x},y)| \leq c_J^{-1} \int_0^1 \left| \omega \left(\frac{x - \gamma(s,y)}{s} \right) - \omega \left(\frac{\bar{x} - \gamma(s,y)}{s} \right) \right| \frac{ds}{s^n}$$
$$+ \int_0^1 \left| \left(\frac{x - \gamma(s,y)}{s} \right) \omega \left(\frac{x - \gamma(s,y)}{s} \right) - \left(\frac{\bar{x} - \gamma(s,y)}{s} \right) \omega \left(\frac{\bar{x} - \gamma(s,y)}{s} \right) \right| \frac{ds}{s^n}$$

But $\omega(z)$ and $z\omega(z)$ are Hölder α on bounded domains for any $0 < \alpha < 1$. Therefore, assuming, for example, $|x - y| \leq |\bar{x} - y|$, and using that the integrand in the definition of $G(x,y)$ vanishes if $s < c|x - y|$ we obtain

$$|G(x,y) - G(\bar{x},y)| \leq C|x - \bar{x}|^\alpha \int_{c|x-y|}^1 \frac{ds}{s^{n+\alpha}} \leq C \frac{|x - \bar{x}|^\alpha}{|x - y|^{n-1+\alpha}}$$

and then,

$$|\mathbf{u}(x) - \mathbf{u}(\bar{x})| \leq C|x - \bar{x}|^\alpha \int_\Omega \frac{|f(y)|}{|x - y|^{n-1+\alpha}} \, dy$$

now, since $p > n$ we can choose $\alpha > 0$ such that $(n - 1 + \alpha)p' < n$ and the proof concludes using the Hölder inequality. $\square$

In what follows we will prove that $\mathbf{u}$ belongs to $W_0^{1,p}(\Omega)$. The argument is analogous to that used in the case of star-shaped domains, i, e., we will write the derivatives of the components of $\mathbf{u}$ as a singular integral operator acting on f. With this goal we introduce

$$T_{ij}g(y) = \lim_{\varepsilon \to 0} \int_{|y-x|>\varepsilon} \chi_\Omega(y) \frac{\partial G_i}{\partial x_j}(x,y) g(x) \, dx.$$

and its adjoint operator T_{ij}^*. The existence of this limit in L^p, for $1 < p < \infty$, will be proved below using the Calderón-Zygmund operator theory. In particular, it will follow that

$$T^*_{ij}f(x) = \lim_{\varepsilon \to 0} \int_{|y-x|>\varepsilon} \chi_\Omega(y) \frac{\partial G_i}{\partial x_j}(x,y) f(y) \, dy$$

Lemma 2.10. *For* $1 \le i,j \le n$ *we have*

$$\frac{\partial u_i}{\partial x_j} = T^*_{ij}f + \omega_{ij}f \quad in \ \ \Omega$$

where

$$\omega_{ij}(y) = \int_{\mathbb{R}^n} \frac{z_i z_j}{|z|^2} \omega(-\dot\gamma(0,y)+z) \, dz$$

Proof. Proceeding as in Lemma 2.3 we have, for $\varphi \in C_0^\infty(\Omega)$ and $y \in \Omega$,

$$\int_\Omega \frac{\partial u_i}{\partial x_j}(x)\varphi(x)\,dx = \int_\Omega I(y) f(y) \, dy \qquad (2.2.12)$$

where

$$I(y) = \lim_{\varepsilon \to 0} \left\{ \int_{|x-y|>\varepsilon} \frac{\partial G_i}{\partial x_j}(x,y)\varphi(x)\,dx - \int_{|y-\zeta|=\varepsilon} G_i(\zeta,y)\varphi(\zeta) \frac{(y_j-\zeta_j)}{|y-\zeta|}\,d\zeta \right\} \qquad (2.2.13)$$

Proceeding as in Lemma 2.3, using now (2.2.9), the surface integral can be written as

$$-\int_{|y-\zeta|=\varepsilon} G_i(\zeta,y)\varphi(\zeta) \frac{(y_j-\zeta_j)}{|y-\zeta|}\,d\zeta = -\varphi(y)A_\varepsilon(y) + O(\varepsilon) \qquad (2.2.14)$$

with

$$A_\varepsilon(y) := \int_{|y-\zeta|=\varepsilon} G_i(\zeta,y) \frac{(y_j-\zeta_j)}{|y-\zeta|}\,d\zeta$$

But, from the definition of G we have

$$A_\varepsilon(y) = \int_{|\zeta-y|=\varepsilon} \int_0^1 \left(\dot\gamma_i(s,y) + \frac{\zeta_i - \gamma_i(s,y)}{s} \right) \omega\left(\frac{\zeta - \gamma(s,y)}{s} \right) \frac{(y_j-\zeta_j)}{|y-\zeta|} \frac{ds}{s^n}\,d\zeta$$

and making the change of variables $r = \varepsilon/s$ we obtain

$$A_\varepsilon(y) = \int_{|\zeta-y|=\varepsilon} \int_\varepsilon^\infty \left(\dot\gamma_i(\varepsilon/r,y) + \frac{\zeta_i - \gamma_i(\varepsilon/r,y)}{\varepsilon/r} \right) \omega\left(\frac{\zeta - \gamma(\varepsilon/r,y)}{\varepsilon/r} \right) \frac{(y_j-\zeta_j)}{|y-\zeta|} \frac{r^{n-2}}{\varepsilon^{n-1}}\,dr\,d\zeta$$

while a further change of variables $\sigma = (\zeta-y)/\varepsilon$ yields

$$A_\varepsilon(y) = -\int_{|\sigma|=1} \int_\varepsilon^\infty \left(\dot\gamma_i(\varepsilon/r,y) + \frac{y_i+\varepsilon\sigma_i - \gamma_i(\varepsilon/r,y)}{\varepsilon/r} \right) \omega\left(\frac{y+\varepsilon\sigma - \gamma(\varepsilon/r,y)}{\varepsilon/r} \right) \sigma_j r^{n-2}\,dr\,d\sigma$$

$$= -\int_{|\sigma|=1} \int_\varepsilon^\infty \left(\dot\gamma_i(\varepsilon/r,y) + \frac{\gamma_i(0,y) - \gamma_i(\varepsilon/r,y)}{\varepsilon/r} + r\sigma_i \right) \omega\left(\frac{\gamma(0,y) - \gamma(\varepsilon/r,y)}{\varepsilon/r} + r\sigma \right) \sigma_j r^{n-2}\,dr\,d\sigma$$

where we have used that $y = \gamma(0, y)$. But, from 4) of Lemma 2.7 we know that $\dot{\gamma}(s, y)$ is continuous at $s = 0$, and therefore, the integrand tends to $\sigma_i \sigma_j \omega(-\dot{\gamma}(0, y) + r\sigma) r^{n-1}$ for $\varepsilon \to 0$. Moreover, recalling that ω has compact support, we can restrict the integral to bounded r, and since the integrand is bounded we can apply the dominated convergence theorem to obtain

$$\lim_{\varepsilon \to 0} A_\varepsilon(y) = -\int_{|\sigma|=1} \int_0^\infty \sigma_i \sigma_j \omega(-\dot{\gamma}(0, y) + r\sigma) r^{n-1} dr d\sigma = -\int_{\mathbb{R}^n} \frac{z_i z_j}{|z|^2} \omega(-\dot{\gamma}(0, y) + z) dz.$$

Therefore, from (2.2.13) and (2.2.14), we conclude that

$$I(y) = T_{ij} \varphi(y) + \omega_{ij}(y) \varphi(y),$$

replacing in (2.2.12) the lemma is proved. $\square$

Now, our goal is to prove the estimate

$$\|\mathbf{u}\|_{W^{1,p}(\Omega)} \leq C \|f\|_{L^p(\Omega)}$$

for $1 < p < \infty$.

In view of Lemma 2.10, and observing that the function ω_{ij} is bounded, our problem reduces to show that T_{ij}^* is a bounded operator in L^p for $1 < p < \infty$.

By duality, it is enough to prove that T_{ij} is bounded. To simplify notation we drop the subscripts i, j from the operator and introduce the function

$$\psi(a, z) = \frac{\partial}{\partial z_j} \big((a + z_i) \omega(z) \big)$$

for $a \in \mathbb{R}$ and $z \in \mathbb{R}^n$. Then, we have to prove the continuity of an operator of the form

$$Tg(y) = \lim_{\varepsilon \to 0} T_\varepsilon g(y) \tag{2.2.15}$$

where, for $\varepsilon > 0$, T_ε is given by

$$T_\varepsilon g(y) = \int_{|x-y|>\varepsilon} K(x, y) g(x) dx \tag{2.2.16}$$

with

$$K(x, y) = \chi_\Omega(y) \int_0^1 \psi \left(\dot{\gamma}_i(s, y), \frac{x - \gamma(s, y)}{s} \right) \frac{ds}{s^{n+1}} \tag{2.2.17}$$

where ψ is a bounded function such that its support in z is contained in that of ω. Since ψ is a derivative of a function with compact support we have

$$\int \psi(a, z) dz = 0. \tag{2.2.18}$$

Moreover, proceeding exactly as in Lemma 2.8 we can prove that

$$|K(x,y)| \leq \frac{C}{|x-y|^n} \tag{2.2.19}$$

with $C = C(n, c_J, \omega)$.

Lemma 2.11. *There exists a constant $C_3 = C_3(n, c_J)$ such that, if $K(x,y) \neq 0$, then*

$$|x-y| \leq C_3 d(x)$$

Proof. Recalling that $\psi(a,z)$ vanishes whenever $z \notin supp\, \omega \subset B(0, c_J/2)$, and using 2) of Lemma 2.7 we know that

$$|x - \gamma(s,y)| \leq \frac{c_J s}{2} \leq \frac{d(\gamma(s,y))}{2} \tag{2.2.20}$$

and so, recalling that $\gamma(0,y) = y$ and that $\dot{\gamma}(s,y) \leq c_J^{-1}$, we obtain

$$|x-y| \leq |x - \gamma(s,y)| + |\gamma(s,y) - \gamma(0,y)| \leq \frac{c_J s}{2} + \sqrt{n} c_J^{-1} s$$

therefore, using again 2) of Lemma 2.7, it follows that

$$|x-y| \leq Cd(\gamma(s,y)) \tag{2.2.21}$$

with $C = C(n, c_J)$. But, the function d is Lipschitz with constant 1 and then, it follows from (2.2.20) that

$$d(\gamma(s,y)) - d(x) \leq |\gamma(s,y)) - x| \leq \frac{1}{2} d(\gamma(s,y))$$

and therefore,

$$d(\gamma(s,y)) \leq 2d(x)$$

which together with (2.2.21) concludes the proof. $\square$

In order to prove the continuity of the operator defined in (2.2.15) and (2.2.16), in the next lemma we decompose it in three parts. The first one will be bounded using Theorem 2.2 while the other two parts using Theorem 2.3.

In view of the previous lemma we have

$$T_\varepsilon g(y) = \int_{\varepsilon < |x-y| \leq C_3 d(x)} K(x,y) g(x) dx.$$

Since for $|x-y| \leq d(x)/2$ we have $d(y)/3 \leq d(x)/2$, and assuming $C_3 > 1/2$, we can decompose the operator as

$$T_\varepsilon g(y) = T_{1,\varepsilon} g(y) + T_2 g(y) + T_3 g(y) \tag{2.2.22}$$

with

$$T_{1,\varepsilon}g(y) = \int_{\varepsilon < |x-y| \leq d(y)/3} K(x,y)g(x)dx, \tag{2.2.23}$$

$$T_2g(y) = \int_{d(y)/3 < |x-y| \leq d(x)/2} K(x,y)g(x)dx$$

and

$$T_3g(y) = \int_{d(x)/2 < |x-y| \leq C_3 d(x)} K(x,y)g(x)dx$$

Lemma 2.12. *For* $1 < p < \infty$, *there exists a constant* C *depending on* C_3, ω, n, *and* p *such that*

$$\|T_2g\|_{L^p} + \|T_3g\|_{L^p} \leq C\|g\|_{L^p} \tag{2.2.24}$$

Proof. To bound T_2 observe that, $|x-y| \leq d(x)/2$ implies $d(x) \leq 2d(y)$, and therefore, using (2.2.19), we have

$$|T_2g(y)| \leq CMg(y), \tag{2.2.25}$$

and therefore, it follows from Theorem 2.3 that T_2 is bounded in L^p.

On the other hand, for any $f \in L^{p'}$, we have

$$\int T_3g(y)f(y)dy = \int\int_{d(x)/2 < |x-y| \leq C_3 d(x)} K(x,y)g(x)f(y)dxdy, \tag{2.2.26}$$

changing the order of integration and using again (2.2.19) we obtain

$$\left| \int T_3g(y)f(y)dy \right| \leq C \int \left\{ \frac{1}{d(x)^n} \int_{|x-y| \leq C_3 d(x)} |f(y)|dy \right\} |g(x)|dx \leq C \int Mf(x)|g(x)|dx$$

which by duality and using again the boundedness of the maximal operator given in Theorem 2.3 gives the bound for T_3. $\square$

Finally, we will show that the singular part $T_{1,\varepsilon}$ can be bounded using Theorems 2.2 and 2.3. With this goal we introduce

$$Sg(y) = \lim_{\varepsilon \to 0} S_\varepsilon g(y) = \lim_{\varepsilon \to 0} \int_{|x-y| > \varepsilon} H(y, x-y)g(x)dx \tag{2.2.27}$$

where

$$H(y,z) = \chi_\Omega(y) \int_0^\infty \psi\left(\dot{\gamma}_t(0,y), \frac{z}{s} - \dot{\gamma}(0,y)\right) \frac{ds}{s^{n+1}}$$

Lemma 2.13. *The kernel* $H(y,z)$ *satisfies the hypotheses of Theorem 2.2.*

Proof. To prove 1), given $\lambda > 0$ we make the change of variable $t = s/\lambda$. Then, we have

$$H(y, \lambda z) = \int_0^\infty \psi\left(\dot{\gamma}_i(0,y), \frac{\lambda z}{s} - \dot{\gamma}(0,y)\right) \frac{ds}{s^{n+1}}$$

$$= \lambda^{-n} \int_0^\infty \psi\left(\dot{\gamma}_i(0,y), \frac{z}{t} - \dot{\gamma}(0,y)\right) \frac{dt}{t^{n+1}} = \lambda^{-n} H(y,z)$$

To prove 2), we make now the change of variable $r = 1/s$ in the integral defining $H(y,z)$ to obtain

$$H(y,z) = \chi_\Omega(y) \int_0^\infty \psi(\dot{\gamma}_i(0,y), rz - \dot{\gamma}(0,y)) \, r^{n-1} dr$$

and therefore,

$$\int_\Sigma H(y,\sigma) d\sigma = \chi_\Omega(y) \int_\Sigma \int_0^\infty \psi(\dot{\gamma}_i(0,y), r\sigma - \dot{\gamma}(0,y)) \, r^{n-1} dr d\sigma$$

$$= \chi_\Omega(y) \int_{\mathbb{R}^n} \psi(\dot{\gamma}_i(0,y), z - \dot{\gamma}(0,y)) \, dz = 0$$

where we have used (2.2.18).

Finally, since the support of ψ in its second variable is contained in $B(0, c_J/2)$ and $|\dot{\gamma}(0,y)| \leq c_J^{-1}$, there exists a constant C depending only on c_J such that

$$|H(y,z)| \leq \int_{|z|/C}^\infty \left| \psi\left(\dot{\gamma}_i(0,y), \frac{z}{s} - \dot{\gamma}(0,y)\right) \right| \frac{ds}{s^{n+1}}$$

which, using that ψ is bounded, implies 3). □

Corollary 2.1. *For $1 < p < \infty$, the operator*

$$Sg(y) = \lim_{\varepsilon \to 0} S_\varepsilon g(y)$$

defined in (2.2.27) is a bounded operator in L^p and the convergence holds in the L^p norm. Moreover, there exists a constant C, depending on p, n, c_J, and ω such that, if

$$\widetilde{S}g(y) = \sup_{\varepsilon > 0} |S_\varepsilon g(y)|,$$

then

$$\left\| \widetilde{S}g \right\|_{L^p} \leq C \|g\|_{L^p}$$

Proof. It follows immediately from Lemma 2.13 and Theorem 2.2. □

Corollary 2.2. *For $1 < p < \infty$, the operator*

$$T_1 g(y) = \lim_{\varepsilon \to 0} T_{1,\varepsilon} g(y),$$

with $T_{1,\varepsilon}$ as in (2.2.23), defines a bounded operator in L^p and the convergence holds in the L^p norm.

Proof. According to 4) in Lemma 2.7, for $0 \leq s \leq kd(y)$, we have $\gamma(s,y) = y + \dot{\gamma}(0,y)s$, and therefore, the kernel defined in (2.2.17) can be written as

$$K(x,y) = \chi_\Omega(y) \left\{ \int_0^{kd(y)} \psi\left(\dot{\gamma}_i(0,y), \frac{x-y}{s} - \dot{\gamma}(0,y) \right) \frac{ds}{s^{n+1}} \right.$$
$$\left. + \int_{kd(y)}^1 \psi\left(\dot{\gamma}_i(s,y), \frac{x-\gamma(s,y)}{s} \right) \frac{ds}{s^{n+1}} \right\}$$

and then, defining

$$J(x,y) = \chi_\Omega(y) \left\{ -\int_{kd(y)}^\infty \psi\left(\dot{\gamma}_i(0,y), \frac{x-y}{s} - \dot{\gamma}(0,y) \right) \frac{ds}{s^{n+1}} \right.$$
$$\left. + \int_{kd(y)}^1 \psi\left(\dot{\gamma}_i(s,y), \frac{x-\gamma(s,y)}{s} \right) \frac{ds}{s^{n+1}} \right\}$$

we obtain

$$K(x,y) = H(y, x-y) + J(x,y)$$

or, in other words,

$$T_{1,\varepsilon}g(y) = \int_{\varepsilon < |x-y| \leq d(y)/3} H(y, x-y)g(x)dx + \int_{\varepsilon < |x-y| \leq d(y)/3} J(x,y)g(x)dx$$

which can be rewritten as

$$T_{1,\varepsilon}g(y) = S_\varepsilon g(y) - \int_{|x-y| > d(y)/3} H(y, x-y)g(x)dx + \int_{\varepsilon < |x-y| \leq d(y)/3} J(x,y)g(x)dx.$$

Now, it is easy to see that there exists a constant, depending only on n, k and the L^∞-norm of ψ such that $|J(x,y)| \leq C/d(y)^n$, and therefore, the third term on the right-hand side is controlled by $Mg(y)$. Then, it follows from Corollary 2.1, that the limit defining $T_1g(y)$ exists in L^p, and moreover,

$$|T_1g(y)| \leq C\left\{ |Sg(y)| + \tilde{S}g(y) + Mg(y) \right\},$$

and we conclude the proof applying again Corollary 2.1 and Theorem 2.3. $\quad\square$

Summing up we obtain our main theorem.

Theorem 2.6. *Let $\Omega \subset \mathbb{R}^n$ be a bounded John domain with constant c_1 with respect to $x_0 = 0$. If $f \in L^p(\Omega)$, $1 < p < \infty$, and $\int_\Omega f = 0$, then the function $\mathbf{u}$ defined in (2.2.10) is in $W_0^{1,p}(\Omega)^n$ and satisfies*

$$\operatorname{div}\mathbf{u} = f \qquad \text{in } \Omega.$$

Moreover, there exists a constant $C = C(c_1, d(x_0), \operatorname{diam}(\Omega), n, p)$ such that

$$\|\mathbf{u}\|_{W^{1,p}(\Omega)^n} \leq C\|f\|_{L^p(\Omega)} \tag{2.2.28}$$

Proof. First, using the bound for G given in (2.2.9) we obtain, by an application of the Young inequality, that $\mathbf{u} \in L^p(\Omega)^n$ and

$$\|\mathbf{u}\|_{L^p(\Omega)^n} \leq C\|f\|_{L^p(\Omega)}. \tag{2.2.29}$$

From Theorem 2.5 we know that div $\mathbf{u} = f$. On the other hand, from Lemma 2.10 we know that

$$\frac{\partial u_i}{\partial x_j} = T_{ij}^* f + \omega_{ij} f \quad \text{in } \Omega.$$

with $\omega_{ij}(y)$ bounded, indeed, $\|\omega_{ij}\|_{L^\infty} \leq \|\omega\|_{L^1}$. Then, (2.2.28) is a consequence of (2.2.29) and the boundedness of T_{ij}^* or, by duality, of T_{ij}. But this follows from (2.2.22), (2.2.24) and Corollary 2.2. In all the estimates used to obtain (2.2.28) the constants depends only on $p, n, c_J, diam(\Omega)$. But, from Lemma 2.7, the constant c_J depend on c_1 and $d(x_0)$.

It only remains to show that $\mathbf{u} \in W_0^{1,p}$. For $p > n$ we have proved in Proposition 2.2 that $\mathbf{u}$ is continuous and vanishes on $\partial\Omega$, and then, as in the case of star-shaped domains, it follows from [85] that $\mathbf{u} \in W_0^{1,p}(\Omega)^n$. For $1 < p \leq n$ we proceed by density as we have done in the proof of Theorem 2.4. $\square$

In some applications it is of interest to have a generalization of Theorem 2.6 to weighted Sobolev spaces. Below we will show that such a result can be obtained as a consequence of a general theorem for singular integral operators for weights in the Muckenhoupt class A_p.

For $1 < p < \infty$, a non-negative function w defined in $\mathbb{R}^n$ is in A_p if

$$\sup_{Q \subset \mathbb{R}^n} \left(\frac{1}{|Q|}\int_Q w\right)\left(\frac{1}{|Q|}\int_Q w^{-1/(p-1)}\right)^{p-1} < \infty, \tag{2.2.30}$$

where the supremum is taken over all cubes with edges parallel to the coordinate axes.

In the following theorems we will use the sharp maximal function defined as

$$M^\# f(x) = \sup_{Q \ni x} \frac{1}{|Q|}\int_Q |f(y) - f_Q|dy$$

and the Fefferman-Stein inequality which says that

$$\|g\|_{L_w^p} \leq C\|M^\# g\|_{L_w^p} \tag{2.2.31}$$

for any $g \in L_w^p$ (see, for example, [34]).

Theorem 2.7. *Given a singular integral operator*

$$Tf(x) = \lim_{\varepsilon \to 0} \int_{|x-y|>\varepsilon} K(x,y)f(y)dy$$

which is continuous in L^p, for $1 < p < \infty$, and such that $K(x,y)$ satisfies

$$|K(x,y) - K(\bar{x},y)| \leq \frac{C|x-\bar{x}|}{|x-y|^{n+1}}, \qquad for \quad |x-y| \geq 2|x-\bar{x}|$$

then, for any $s > 1$,

$$M^{\#}Tf(x) \leq C(M|f|^s(x))^{1/s}$$

Proof. This estimate is well known and its proof can be found in several books, although the hypotheses on the operator are not stated usually as we are doing here. However, it is easy to check that the proof given in [34, Lemma 7.9] only uses the hypotheses given above. □

Theorem 2.8. *Under the hypotheses of Theorem 2.6, if $w \in A_p$, there exists a constant $C = C(c_1, d(x_0), diam(\Omega), n, p, w)$ such that*

$$\|\mathbf{u}\|_{W^{1,p}(\Omega,w)^n} \leq C\|f\|_{L^p(\Omega,w)} \tag{2.2.32}$$

Proof. It is enough to bound $\partial u_i / \partial x_j$. The estimate for u will follow from the Poincaré inequality, which is known to hold for A_p weights (see, for example, [33]).

Now, using Lemma 2.10 and that $\omega_{i,j}$ is bounded, we have to prove that

$$\|T^* f\|_{W^{1,p}_w(\Omega)^n} \leq C\|f\|_{L^p_w(\Omega)}$$

where

$$T^* f(x) = \lim_{\varepsilon \to 0} \int_{|x-y|>\varepsilon} K(x,y)f(y)dy$$

with $K(x,y)$ given in (2.2.17). Proceeding again as in Lemma 2.8 we obtain

$$|\nabla_x K(x,y)| \leq \frac{C}{|x-y|^{n+1}},$$

and consequently,

$$|K(x,y) - K(\bar{x},y)| \leq \frac{C|x-\bar{x}|}{|x-y|^{n+1}}, \qquad for \quad |x-y| \geq 2|x-\bar{x}|,$$

therefore, since we already know that T^* is a continuous operator in L^p, for $1 < p < \infty$, T^* satisfies the hypotheses of Theorem 2.7. Consequently, we have

$$M^{\#}T^* f(x) \leq C(M|f|^s(x))^{1/s}$$

for any $s > 1$.

Now the result follows from this estimate combined with (2.2.31) for $g = T^* f$. We omit details and refer the reader to the proof of Theorem 7.11 in [34]. □

2.3 Improved Poincaré Inequality and Equivalences

This section deals with the so-called improved Poincaré inequality, namely, for $f \in L_0^p(\Omega)$,

$$\|f\|_{L^p(\Omega)} \leq C\|d\nabla f\|_{W^{1,p}(\Omega)^n} \tag{2.3.1}$$

where the constant C depends only on the domain Ω.

This inequality has many applications. In fact, we will show below that it provides a different way to prove the existence of solutions of the divergence in Sobolev spaces. Moreover, we will see in Chapter 3 that it can be used to prove Korn type inequalities.

The next theorem is a particular case of results given in [33]. In its proof we will use the following well-known result.

Lemma 2.14. *For $\beta > 0$ and $g \in L_{loc}^1$,*

$$\int_{|x-y| \leq \beta} \frac{|g(y)|}{|x-y|^{n-1}} dy \leq C\beta Mg(x)$$

with a constant C independent of β and g.

Proof.

$$\int_{|x-y| \leq \beta} \frac{|g(y)|}{|x-y|^{n-1}} dy \leq \sum_{k=0}^{\infty} \int_{2^{-(k+1)}\beta < |x-y| \leq 2^{-k}\beta} \frac{|g(y)|}{|x-y|^{n-1}} dy$$

$$\leq C\beta \sum_{k=0}^{\infty} \frac{2^{-k}}{|B(x, 2^{-k}\beta)|} \int_{|x-y| \leq 2^{-k}\beta} |g(y)| dy \leq C\beta Mg(x)$$

$\square$

Theorem 2.9. *If $\Omega \subset \mathbb{R}^n$ is a bounded John domain, then the improved Poincaré inequality (2.3.1) holds for $1 \leq p < \infty$.*

Proof. By Lemma 3.1 it is enough to prove

$$\|f - f_\omega\|_{L^p(\Omega)} \leq C\|d\nabla f\|_{L^p(\Omega)^n}$$

with ω is as in the previous section.

By density we can assume that $f \in C^1(\Omega)$. Indeed, the density of $C^1(\Omega)$ in $W^{1,p}(\Omega, 1, d^p)$ can be proved by the same argument used in the unweighted case (see, for example, [40]).

As in Lemma 2.9 we can show that

$$f(y) - f_\omega = -\int_\Omega G(x,y) \cdot \nabla f(x) dx.$$

with $G(x,y)$ given in (2.2.8). Then, given $g \in L^{p'}(\Omega)$, we have

$$\int_\Omega (f(y) - f_\omega) g(y) dy = \int_\Omega \int_\Omega G(x,y) \cdot \nabla f(x) g(y) dx dy$$

Interchanging the order of integration and using Lemmas 2.8 and 2.11, we obtain

$$\int_\Omega (f(y) - f_\omega) g(y) dy \leq C \int_\Omega \left\{ \int_{|x-y| \leq C_3 d(x)} \frac{|g(y)|}{|x-y|^{n-1}} dy \right\} |\nabla f(x)| dx,$$

and therefore, using Lemma 2.14,

$$\int_\Omega (f(y) - f_\omega) g(y) dy \leq C \int_\Omega Mg(x) d(x) |\nabla f(x)| dx \leq C \|Mg\|_{L^{p'}(\Omega)} \|d\nabla f\|_{L^p(\Omega)}$$

and the proof concludes using the continuity of the maximal operator in $L^{p'}$ and duality. $\square$

Remark 2.6. Note that the previous theorem includes the limit case $p = 1$. Clearly, the argument does not apply to $p = \infty$ because it would require the continuity of the maximal operator in L^1. Moreover, the improved Poincaré is not valid in L^∞, indeed, an easy counterexample can be given taking $\Omega = (0,1)$ and $f(x) = \log x$.

The improved Poincaré can be used to prove a decomposition of functions with vanishing integral as a sum of locally supported functions with the same property, actually, it is equivalent to the existence of this decomposition. This decomposition is useful to obtain global from local results, more precisely, to extend to very general domains results which are known for cubes.

In the next theorem we analyze the relation between the existence of solutions of the divergence, the improved Poincaré inequality and the decomposition of functions. The arguments are contained in [35, 36, 60].

We will make use of the Whitney decomposition introduced in (2.2.2) denoting now Q_j, $j \in \mathbb{N}$, the cubes. It is known (see, for example, [84]) that there exists a family of functions $\phi_j \in C_0^\infty(Q_j^*)$ associated with the decomposition such that $\sum_j \phi_j = \chi_\Omega$, $\|\phi_j\|_{L^\infty} \leq C$ and $\|\nabla \phi_j\|_{L^\infty} \leq C/d_j$, where d_j denotes the distance of Q_j to the boundary of Ω.

Theorem 2.10. *Let $\Omega \subset \mathbb{R}^n$ be an arbitrary domain and $1 < p < \infty$. Consider the following statements,*

1. $\|f\|_{L^{p'}(\Omega)} \leq C \|d\nabla f\|_{L^{p'}(\Omega)^n} \qquad \forall f \in L_0^{p'}(\Omega)$

2. $\forall f \in L_0^p(\Omega)$, *there exists* $\mathbf{u} \in L^p(\Omega)^n$ *such that*

$$div\, \mathbf{u} = f \quad in\ \Omega\ ,\ \mathbf{u} \cdot n = 0 \quad on\ \partial\Omega$$

and

$$\left\| \frac{\mathbf{u}}{d} \right\|_{L^p(\Omega)^n} \leq C \|f\|_{L^p(\Omega)}$$

3. $\forall f \in L_0^p(\Omega)$, there exists a decomposition

$$f = \sum_j f_j$$

such that

$$f_j \in L_0^p(\Omega), \quad \operatorname{supp} f_j \subset Q_j^*, \quad and \quad \|f\|_{L^p(\Omega)}^p \sim \sum_j \|f_j\|_{L^p(Q_j^*)}^p$$

4. $\forall f \in L_0^p(\Omega)$, there exists $\mathbf{u} \in W_0^{1,p}(\Omega)^n$ such that

$$div\, \mathbf{u} = f \quad in\; \Omega$$

and

$$\|D\mathbf{u}\|_{L^p(\Omega)^n} \le C\|f\|_{L^p(\Omega)}$$

Then,

$$(1) \iff (2) \iff (3) \implies (4)$$

and the constants are equivalent, i.e., the ratio between two of them is bounded by above and below by positive constants depending only on n and p.

Proof. $(1) \Rightarrow (2)$: For $f \in L_0^p(\Omega)$

$$\mathscr{L}(\nabla g) = \int_\Omega f g$$

defines a linear form on the subspace of $L^{p'}(\Omega)^n$ formed by the gradient vector fields. Note that $\mathscr{L}$ is well defined because $\int_\Omega f = 0$. Moreover, it follows from (1) that

$$|\mathscr{L}(\nabla g)| = \left| \int_\Omega f(g - g_\Omega) \right| \le C\|f\|_{L^p(\Omega)} \|d\nabla g\|_{L^{p'}(\Omega)^n}.$$

By the Hahn-Banach theorem $\mathscr{L}$ can be extended as a linear continuous functional

$$\mathscr{L} : L^{p'}(\Omega, d^{p'})^n \longrightarrow \mathbb{R}$$

and therefore, by duality, there exists $\mathbf{u} \in L^p(\Omega, d^{-p})^n$ such that

$$\mathscr{L}(\mathbf{v}) = \int_\Omega \mathbf{u} \cdot \mathbf{v} \quad and \quad \left\| \frac{\mathbf{u}}{d} \right\|_{L^p(\Omega)^n} \le C\|f\|_{L^p(\Omega)}$$

in particular,

$$\int_\Omega \mathbf{u} \cdot \nabla g = \int_\Omega f g \quad \forall g \in W^{1,p'}(\Omega)$$

which is equivalent to

$$div\, \mathbf{u} = f \quad in\; \Omega, \; \mathbf{u} \cdot n = 0 \quad on\; \partial\Omega$$

and so (2) holds.

$(2) \Rightarrow (1)$: Given $f \in L_0^p(\Omega)$ we have

$$\|f\|_{L^{p'}(\Omega)} = \sup_{g \in L_0^p(\Omega), \|g\|_p = 1} \int_\Omega fg \qquad (2.3.2)$$

Now, for $g \in L_0^p(\Omega)$, let $\mathbf{u} \in L^p(\Omega)^n$ be the solution of div $\mathbf{u} = g$ given by (2). Then,

$$\int_\Omega fg = \int_\Omega f \operatorname{div} \mathbf{u} = \int_\Omega \nabla f \cdot \mathbf{u}$$

$$\leq \|d\nabla f\|_{L^{p'}(\Omega)^n} \left\|\frac{\mathbf{u}}{d}\right\|_{L^p(\Omega)} \leq C \|d\nabla f\|_{L^{p'}(\Omega)^n} \|g\|_{L^p(\Omega)}$$

which together with (2.3.2) implies (1).

$(2) \Rightarrow (3)$: Given $f \in L_0^p(\Omega)$ let $\mathbf{u} \in L^p(\Omega)^n$ given by (2). We define

$$f_j = \operatorname{div}(\phi_j \mathbf{u})$$

then

$$f = \operatorname{div} \mathbf{u} = \operatorname{div}\left(\mathbf{u} \sum_j \phi_j\right) = \sum_j \operatorname{div}(\phi_j \mathbf{u}) = \sum_j f_j$$

Since $\operatorname{supp} \phi_j \subset Q_j^*$ we have $\operatorname{supp} f_j \subset Q_j^*$ and $\int_\Omega f_j = 0$.

Moreover, from the finite superposition of the Whitney decomposition we have

$$|f(x)|^p \leq C \sum_j |f_j(x)|^p$$

and then

$$\|f\|_{L^p(\Omega)}^p \leq C \sum_j \|f_j\|_{L^p(Q_j^*)}^p$$

where the constant C depends only on p and n.

To prove the other inequality we use again the finite superposition and that $\|\phi_j\|_{L^\infty} \leq 1$ and $\|\nabla \phi_j\|_{L^\infty} \leq C/d_j$. Then, we have

$$\|f_j\|_{L^p(Q_j^*)}^p \leq C \left\{ \|f\|_{L^p(Q_j^*)}^p + \left\|\frac{\mathbf{u}}{d}\right\|_{L^p(Q_j^*)^n}^p \right\}$$

and therefore it follows from (2) that

$$\sum_j \|f_j\|_{L^p(Q_j^*)}^p \leq C \|f\|_{L^p(\Omega)}^p$$

$(3) \Rightarrow (2)$: Given $f \in L_0^p(\Omega)$ we write $f = \sum_j f_j$ according to (3). From the results in Section 2.1 we know that, for each j, there exists $\mathbf{u}_j \in W_0^{1,p}(Q_j^*)^n$ such that

$$\operatorname{div} \mathbf{u}_j = f_j \quad \text{and} \quad \|D\mathbf{u}_j\|_{L^p(Q_j^*)} \leq C \|f_j\|_{L^p(Q_j^*)}$$

where the constant is independent of the size of the cube. Then, $\mathbf{u} = \sum_j \mathbf{u}_j \in W_0^{1,p}(\Omega)^n$ is the required $\mathbf{u}$. Indeed, $\operatorname{div} \mathbf{u} = f$ and the estimate

$$\left\|\frac{\mathbf{u}}{d}\right\|_{L^p(\Omega)^n} \le C \|f\|_{L^p(\Omega)}$$

follows applying the Poincaré inequality on each Q_j^*. Indeed

$$\left\|\frac{\mathbf{u}_j}{d}\right\|_{L^p(Q_j^*)^n} \sim \frac{1}{d_j} \|\mathbf{u}_j\|_{L^p(Q_j^*)^n} \le C \|D\mathbf{u}_j\|_{L^p(Q_j^*)^n} \le C \|f_j\|_{L^p(Q_j^*)}$$

then

$$\left\|\frac{\mathbf{u}}{d}\right\|_{L^p(Q_j^*)^n}^p \le C \sum_j \left\|\frac{\mathbf{u}_j}{d}\right\|_{L^p(Q_j^*)^n}^p \le C \sum_j \|f_j\|_{L^p(Q_j^*)}^p \le C \|f\|_{L^p(\Omega)}^p$$

as we wanted to show.

(3) $\Rightarrow$ (4): It is proved exactly as (3) $\Rightarrow$ (2).

The equivalence between the constants follows easily from the proofs. $\square$

As we mentioned above, the results given in this section provide a different argument to prove the existence of solutions of the divergence. This is summarized in the following corollary. The same argument has been used in [32]

Corollary 2.3. *If* $\Omega \subset \mathbb{R}^n$ *is a bounded John domain and* $1 \le p < \infty$ *then for any* $f \in L_0^p(\Omega)$ *there exists* $\mathbf{u} \in W_0^{1,p}(\Omega)^n$ *such that*

$$\operatorname{div} \mathbf{u} = f \quad \text{in } \Omega$$

and

$$\|\mathbf{u}\|_{W^{1,p}(\Omega)^n} \le C \|f\|_{L^p(\Omega)}$$

Proof. It is an immediate consequence of Theorems 2.9 and 2.10. $\square$

Remark 2.7. We don't know whether (4) implies the other statements given in Theorem 2.10 for a general domain. However, it is easy to see that (4) $\Rightarrow$ (2) holds for domains satisfying the Hardy inequality, i.e., there exists a constant depending only on Ω and p such that

$$\left\|\frac{v}{d}\right\|_{L^p(\Omega)} \le C \|\nabla v\|_{L^p(\Omega)^n} \qquad \forall v \in W_0^{1,p}(\Omega)$$

Moreover, it is known that this inequality is valid for any domain different from $\mathbb{R}^n$ when $p > n$ [73]. Therefore, for $p > n$ we have the stronger statement

$$(1) \iff (2) \iff (3) \iff (4)$$

for any domain $\Omega \not\subseteq \mathbb{R}^n$.

2.4 A Partial Converse Result

An interesting problem is to characterize the bounded domains for which the results considered in Theorem 2.10 are valid. According to the previous section we know that all of them hold for John domains. As we have mentioned at the beginning of Section 2.2, it is known that div_p is not valid for some domains with external cusps. Since the class of John domains is very general and excludes external cusps it seems a natural question whether

$$\Omega \quad \text{satisfies} \quad div_p \iff \Omega \quad \text{is a John domain.} \qquad (2.4.1)$$

As far as we know the answer is not known. However, a partial answer can be given. Indeed, (2.4.1) is true if the bounded domain Ω satisfies the separation property. We omit the technical definition of this property and refer the reader to [16] where it was introduced. In that paper it is also proved that, in the two dimensional case, any simply connected domain satisfies the separation property.

For $1 \leq p < n$ we say that Ω satisfies the Sobolev-Poincaré inequality for p if there exists a constant depending only on p and Ω such that

$$\|f\|_{L^{p^*}(\Omega)} \leq C\|\nabla f\|_{L^p(\Omega)^n} \qquad \forall f \in W^{1,p}(\Omega) \cap L_0^p(\Omega),$$

where $p^* = pn/(n-p)$. In [16] the authors prove that, if Ω is a bounded domain that satisfies the separation property as well as the Sobolev-Poincaré inequality for some $1 < p < n$, then it is a John domain.

Theorem 2.11. *Let $\Omega \subset \mathbb{R}^n$ be a bounded domain satisfying the separation property. Then, Ω satisfies div_p, for some $1 < p < \infty$, if and only if Ω is a John domain.*

Proof. From the previous section we already know that if Ω is a John domain then div_p is valid for all $1 < p < \infty$. The converse was proved in [3] in the case $1 < p < n$ showing that div_p implies the Sobolev-Poincaré for $q = (p^*)'$ and applying the result in [16]. Indeed, given $f \in W^{1,q}(\Omega) \cap L_0^q(\Omega)$ and $g \in L_0^p(\Omega)$, take $\mathbf{u} \in W_0^{1,p}(\Omega)^n$ such that $\text{div}\,\mathbf{u} = g$ and $\|\mathbf{u}\|_{W^{1,p}(\Omega)^n} \leq C\|g\|_{L^p(\Omega)}$. Now, by the Sobolev-Poincaré for functions in $W_0^{1,p}(\Omega)^n$ and observing that $q' = p^*$ we have

$$\int_\Omega fg = \int_\Omega f\text{div}\,\mathbf{u} = -\int_\Omega \nabla f \cdot \mathbf{u} \leq \|\nabla f\|_{L^q(\Omega)^n}\|\mathbf{u}\|_{L^{q'}(\Omega)^n}$$
$$\leq C\|\nabla f\|_{L^q(\Omega)^n}\|\mathbf{u}\|_{W^{1,p}(\Omega)^n} \leq C\|\nabla f\|_{L^q(\Omega)^n}\|g\|_{L^p(\Omega)}$$

and the argument concludes observing that $p = (q^*)'$ and using duality.

For the case $n < p$ the result was proved in [60] generalizing the arguments of [16] to show that, under the separation property, the improved Poincaré inequality implies that Ω is a John domain, and then using that div_p implies the improved Poincaré for p' (see Theorem 2.10 and Remark 2.7). We can also use the following argument: in [35] it is proved that, if the improved Poincaré is valid for some $q \geq 1$ then it is valid for all r such that $q \leq r < \infty$, actually, in that paper the proof is written for $q = 1$ but the argument can be easily extended to any q. Now, assuming div_p for some $p > n$ we have that Ω satisfies the improved Poincaré for p', and therefore, for p. Then, using again Theorem 2.10, we obtain that Ω satisfies $div_{p'}$. But $p' < n$, and so, Ω is a John domain. Finally, the case $p = n$ was proved in [59]. $\square$

2.5 Comments and References

An interesting problem that has been widely considered is that of the dependence on the domain of the constants involved in all the inequalities considered. Ideally, given a particular Ω and an inequality, one would like to know the best constant possible, but this is a too difficult problem that can be solved only in very particular cases. For example, the constant for div_2 is known for circles, ellipses, spheres, and spherical shells. We refer the reader to [56] and the references therein.

A less ambitious problem is to obtain estimates of the constants in terms of geometric properties of the domains. There are many works in this direction. Important tools in this problem are results like Theorem 2.10 which allow to translate information for some inequality to another one.

Consider, for example, the estimate

$$\|D\mathbf{u}\|_{L^p(\Omega)} \le C_{p,div,\Omega} \|f\|_{L^p(\Omega)}, \tag{2.5.1}$$

where $\mathbf{u} \in W_0^{1,p}(\Omega)^n$ is some solution of $\operatorname{div} \mathbf{u} = f$.

One could try to obtain information tracing constants in the proofs given in the previous sections for the estimates for the solutions of the divergence defined there. However, the arguments are based on the general Calderón-Zygmund singular integral operators theory and the constant that one obtains from that theory seems to be nonoptimal for our particular case. As it was pointed out in [46], for the case of a domain of diameter d which is star-shaped with respect to a ball B of radius ρ, it follows from [21] that, for $1 < p < \infty$, $C_{p,div,\Omega} \le C_{n,p}(d/\rho)^{n+1}$.

However, at least in the case $p = 2$, this estimate can be improved. Indeed, this has been done in [36] where the result given in Section 2.1 is proved for the case $p = 2$ using a different argument. Instead of relying on the general theory, the proof in [36] is based on elementary properties of the Fourier transform. In this way it is proved that, for the solution $\mathbf{u}$ defined in Section 2.1,

$$C_{2,div,\Omega} \le C_n \frac{d}{\rho} \left(\frac{|\Omega|}{|B|} \right)^{\frac{n-2}{2(n-1)}} \left(\log \frac{|\Omega|}{|B|} \right)^{\frac{n}{2(n-1)}} \tag{2.5.2}$$

In particular, in the two dimensional case we have, for any $\varepsilon > 0$,

$$C_{2,div,\Omega} \le C_\varepsilon (d/\rho)^{1+\varepsilon}$$

This estimate has been improved in [28] removing the ε and obtaining

$$C_{2,div,\Omega} \le C(d/\rho) \tag{2.5.3}$$

The result in that paper is not for the solution analyzed in Section 2.1 but for a different one. The authors use the equivalence between $C_{2,div,\Omega}$ and the constant in the so-called Friedrichs inequality. Let h and g be real valued functions such that $\int_\Omega h = 0$ and $h + ig$ is an holomorphic function in Ω. Under appropriate assumptions on the domain Ω, Friedrichs proved that there exists a constant $C_{fr,\Omega}$ such that

$$\|h\|_{L^2(\Omega)} \le C_{fr,\Omega}\|g\|_{L^2(\Omega)}, \tag{2.5.4}$$

Assuming that Ω is a smooth domain it was proved in [57] that, if $C_{2,div,\Omega}$ and $C_{fr,\Omega}$ are the best possible constants in (2.5.1) (with $p = 2$) and (2.5.4), respectively, then

$$C_{2,div,\Omega}^2 = C_{fr,\Omega}^2 + 1$$

This result was extended for arbitrary bounded domains in [28], and using this equivalence and complex variable arguments the authors proved (2.5.3).

This result is optimal, indeed, consider the rectangular domain $\Omega_{a,\varepsilon} = (-a, +a) \times (-\varepsilon, \varepsilon)$, with a and ε positive constants, and take $h(x_1, x_2) = x_1$ and $g(x_1, x_2) = x_2$. Then, an elementary computation shows that the Friedrichs inequality applied to these functions gives

$$C_{fr,\Omega_{a,\varepsilon}} \ge (a/\varepsilon)$$

and consequently, for these domains,

$$C_{2,div,\Omega} \ge c_1(d/\rho) \tag{2.5.5}$$

where c_1 is a constant independent of Ω.

We do not know whether (2.5.2) can be improved for $n > 3$ nor whether similar estimates can be proved for $p \ne 2$.

For the particular case of convex domains it is possible to use $(1) \Rightarrow (4)$ in Theorem 2.10 to prove, for $1 < p < \infty$ and arbitrary dimension n, that

$$C_{p,div,\Omega} \le C(d/\rho)$$

with C depending only on n and p. This has been done in [36] for $p = 2$ but the arguments there can be easily extended to $1 < p < \infty$.

To end our comments on the constant in (2.5.1) let us mention the papers [27] and [11] where the behavior of the constant for domains with corners and continuity with respect to the domain were analyzed.

Finally, several papers have considered the existence of solutions of the divergence in higher order Sobolev spaces under appropriate assumptions on the right-hand side f (see [9, 83, 29]).

Chapter 3
Korn's Inequalities

Introduced in the beginning of the past century [67, 69, 68], the Korn inequality

$$\|\mathbf{Du}\|_{L^2(\Omega)^{n\times n}} \leq C\|\boldsymbol{\varepsilon}(\mathbf{u})\|_{L^2(\Omega)^{n\times n}}, \tag{3.0.1}$$

has become a standard topic in the literature of continuum mechanics. In *elasticity theory* $\mathbf{u}$ plays the role of the displacement field of an elastic body and (3.0.1) states the striking fact that $\mathbf{Du}$ is bounded by its symmetric part. In the jargon of continuum mechanics $\boldsymbol{\varepsilon}(\mathbf{u})$ is called the *linearized strain tensor*. Hooke's law, valid for small deformations, states that stresses depend linearly on $\boldsymbol{\varepsilon}(\mathbf{u})$. In particular, for homogeneous isotropic materials, one can write the stress tensor as follows:

$$\sigma := \lambda tr(\boldsymbol{\varepsilon}(\mathbf{u}))\mathbf{I} + 2\mu\boldsymbol{\varepsilon}(\mathbf{u}) = \lambda \operatorname{div}(\mathbf{u})\mathbf{I} + 2\mu\boldsymbol{\varepsilon}(\mathbf{u}),$$

being λ and μ - the Lamé coefficients - positive parameters empirically obtained for each specific material. Thanks to this constitutive relation, the equations of linear elasticity -for a material body in static equilibrium- can be written in terms of the displacement field

$$- \operatorname{Div} \sigma = -\lambda \nabla \operatorname{div}(\mathbf{u})\mathbf{I} - 2\mu \operatorname{Div} \boldsymbol{\varepsilon}(\mathbf{u}) = \mathbf{g}, \tag{3.0.2}$$

where $\mathbf{g}$ represents the external volumetric forces acting on the body.

If we consider a body occupying a region Ω, for any $\mathbf{g} \in H^{-1}(\Omega)^n$, a solution of (3.0.2) vanishing on the boundary can be sought in the space $\mathbf{V} = H_0^1(\Omega)^n$. Actually, using the orthogonality between symmetric and skew-symmetric matrices in the scalar product $\mathbf{M} : \mathbf{N} = \sum_{i,j=1}^{n} m_{i,j} n_{i,j} = tr(\mathbf{MN}^t)$, we easily get the following weak version of (3.0.2):
Find $\mathbf{u} \in H_0^1(\Omega)^n$ such that

$$\lambda \int_{\Omega} \operatorname{div} \mathbf{u} \operatorname{div} \mathbf{v}\, dx + 2\mu \int_{\Omega} \boldsymbol{\varepsilon}(\mathbf{u}) : \boldsymbol{\varepsilon}(\mathbf{v})\, dx = \langle \mathbf{g}, \mathbf{v} \rangle \qquad \forall \mathbf{v} \in H_0^1(\Omega)^n.$$

Notice that (3.0.1) together with Poincaré inequality for functions with vanishing trace implies the coercivity of the bilinear form associated to this weak formulation.

© The Author(s) 2017
G. Acosta, R.G. Durán, *Divergence Operator and Related Inequalities*,
SpringerBriefs in Mathematics, DOI 10.1007/978-1-4939-6985-2_3

In the first part of this chapter we recall standard versions of Korn's inequalities focusing on their mutual relationship. Some considerations rely on compactness arguments or on some of their implications such as the Poincaré inequality, which is expressed in several equivalent forms that are useful for further discussions. Then we assume that Ω is a John domain and give three different proofs of Korn's inequality in that case.

3.1 Classical Versions of Korn's Inequalities

Notice that both sides of inequality (3.0.1) are well defined for any $\mathbf{u} \in H^1(\Omega)^n$, however, considering any non-constant vector field $\mathbf{w}$ such that $\boldsymbol{\varepsilon}(\mathbf{w}) = 0$ it can be concluded that (3.0.1) does not hold for arbitrary vector fields in $H^1(\Omega)^n$. As a consequence, Korn's inequality in the form (3.0.1) is usually considered under some extra conditions. These conditions have physical roots most of the times and are related to appropriate boundary conditions that guarantee uniqueness of solutions in the elasticity framework.[1] Two classical cases, related to the *pure displacement* or *pure traction* problem of linear elasticity, were early treated in the pioneering works by Korn. These are called, respectively, the *first* and the *second case* of the inequality. They state that (3.0.1) holds if either

- $\mathbf{u} \in H_0^1(\Omega)^n$ (*first case*) or
-

$$\int_{\Omega} \boldsymbol{\mu}(\mathbf{u}) = 0 \qquad (\textit{second case}), \tag{3.1.1}$$

where $\boldsymbol{\mu}(\mathbf{u}) := \textit{skew}(\mathbf{Du})$.

The first case can be proved by means of very simple arguments that hold for any domain Ω. Indeed, consider $\mathbf{u} \in C_0^\infty(\Omega)$. Taking into account that

$$2\mathrm{Div}\,\boldsymbol{\varepsilon}(\mathbf{u}) = \Delta\mathbf{u} + \nabla(\mathrm{div}\,\mathbf{u}),$$

we get after multiplying by $\mathbf{u}$ and integrating by parts

$$2\|\boldsymbol{\varepsilon}(\mathbf{u})\|^2_{L^2(\Omega)^{n\times n}} = \|\mathbf{Du}\|^2_{L^2(\Omega)^{n\times n}} + \|\mathrm{div}\,\mathbf{u}\|^2_{L^2(\Omega)^{n\times n}}.$$

Taking into account the density of $C_0^\infty(\Omega)$ in $H_0^1(\Omega)^n$, Korn's inequality in the first case holds with an universal constant $\sqrt{2}$.

At this point one might wonder if the first case can be extended to L^p. That is, whether

$$\|\mathbf{Du}\|_{L^p(\Omega)^{n\times n}} \leq C\|\boldsymbol{\varepsilon}(\mathbf{u})\|_{L^p(\Omega)^{n\times n}}, \tag{3.1.2}$$

---

[1] Loosely speaking, the goal is to work in a proper subspace of $H^1(\Omega)^n$ for which the condition $\boldsymbol{\varepsilon}(\mathbf{u}) = 0$ implies $\mathbf{u} = 0$.

holds for any $\mathbf{u} \in W_0^{1,p}(\Omega)^n$. Although the argument just used for $p = 2$ does not apply in this context, different approaches show that the answer is still positive. For instance, in [64] this fact is proved, following [86], extending $\mathbf{u}$ by zero outside Ω and using continuity properties of the Riesz transform in $L^p(\mathbb{R}^n)$. In the end of this section we devote some words to this and to the related case in which the field $\mathbf{u}$ vanishes only on a portion $\Gamma \subset \partial\Omega$.

The invariance of (3.1.2) by scalings and the fact that functions in $u \in W_0^{1,p}(\Omega)^n$ can be extended by zero allow to reduce the problem (3.1.2) set on an arbitrary Ω to the same problem on a single fixed domain (for instance, a unitary ball). This, in turn, guarantees the existence of a universal constant in (3.1.2). On the other hand, as we show below, the second case is not valid in arbitrary domains while its proof, even for $p = 2$, requires much deeper considerations.

Since the arguments used in the sequel apply on L^p spaces for any $1 < p < \infty$, we work most of the time with (3.1.2) as well as with the L^p version of (1.0.4), that takes the form

$$\|\mathbf{Du}\|_{L^p(\Omega)^{n\times n}} \leq C\Big\{ \|\mathbf{u}\|_{L^p(\Omega)^n} + \|\boldsymbol{\varepsilon}(\mathbf{u})\|_{L^p(\Omega)^{n\times n}} \Big\}, \qquad (3.1.3)$$

and in which no extra conditions other than $\mathbf{u} \in W^{1,p}(\Omega)^n$ are required.

In this context the *second case* of Korn's inequality, also called the second Korn inequality, involves (3.1.2) and (3.1.1). If it holds, we shortly say that Ω *satisfies* $Korn_p$ or simply, if Ω is understood from the context, that $Korn_p$ holds. Similarly, we call (3.1.3) the *unconstrained case* of Korn's inequality and we say that Ω satisfies $Korn_p^u$ or just that $Korn_p^u$ holds.

The kernel of the operator $\boldsymbol{\varepsilon}$, called the space of *infinitesimal rigid movements*,[2]

$$RM(\Omega)^n = \{ \mathbf{v} \in W^{1,p}(\Omega)^n : \quad \boldsymbol{\varepsilon}(\mathbf{v}) = 0 \},$$

plays a particular role in the following. Using (1.0.6), it is simple to show that every function in RM can be written as $\mathbf{v}(\mathbf{x}) = \mathbf{a} + \mathbf{Mx}$, where $\mathbf{M} \in \mathbb{R}^{n\times n}$ is a skew symmetric matrix.

Consider now

$$\mathscr{S} = \Big\{ \mathbf{w} \in W^{1,p}(\Omega)^n : \quad \boldsymbol{\mu}(\mathbf{w})_\Omega = 0 \Big\},$$

which is a closed subspace of $W^{1,p}(\Omega)^n$.

Notice that $\mathscr{S} \cap RM$ agrees with *the space of constant vector fields*. For any space $\mathscr{W}$ containing constant functions we denote the quotient $[\mathscr{W}] := \mathscr{W}/\mathbb{R}$ with the natural norm

$$\|u\|_{[\mathscr{W}]} := \inf_c \|u - c\|_{\mathscr{W}}$$

and sometimes we identify $u \in \mathscr{W}$ with the class $[u]$ in order to simplify notation.

Let us recall the following,

---

[2] "Real" rigid deformations or movements can be regarded as translations followed by rotations. Accordingly, they are associated to linear mappings defined by means of proper orthogonal matrices.

Definition 3.1.1 (Poincaré Domain) *Let Ω be a domain, ω_1 and ω_2 two weights, with ω_1 integrable. We say that Ω is a Poincaré$-(\omega_1, \omega_2)$ domain if*

$$\|u - u_{\omega_1}\|_{L^p(\Omega, \omega_1)} \leq C \|\nabla u\|_{L^p(\Omega, \omega_2)}$$

for any $u \in W^{1,p}(\Omega, \omega_1, \omega_2)$ and any p. If $\omega_1 = \omega_2 = 1$, then we simply say that Ω is a Poincaré domain.

Remark 3.1. In order to keep the notation as simple as possible our definition only takes into account domains for which the inequality holds for any p. Let us recall that for a given domain the inequality could hold only for p in a certain range.

Lemma 3.1. *Consider $u \in L^p(\Omega, \omega)$ where ω is an integrable weight and choose a bounded function φ such that $\int_\Omega \varphi \omega = 1$.*

1. *The following expressions are equivalent up to a multiplicative constant depending only on p, φ, and ω but not on u*

$$\inf_c \|u - c\|_{L^p(\Omega, \omega)} := \|u\|_{[L^p(\Omega, \omega)]} \sim \|u - u_{\varphi\omega}\|_{L^p(\Omega, \omega)} \sim \|u - u_\omega\|_{L^p(\Omega, \omega)}.$$

2. *Moreover, if $\omega^{-\frac{1}{p-1}}$ is integrable, regular averages can be used in (1), that is*

$$\inf_c \|u - c\|_{L^p(\Omega, \omega)} \sim \|u - u_\varphi\|_{L^p(\Omega, \omega)} \sim \|u - u_\Omega\|_{L^p(\Omega, \omega)}.$$

Proof. (1) We assume $\omega = 1$, the general case follows in the same fashion.

First notice that we have the obvious inequality $\|u\|_{[L^p(\Omega)]} \leq \|u - u_\varphi\|_{L^p(\Omega)}$. Now we write

$$\|u - u_\varphi\|_{L^p(\Omega)} \leq \|u - u_\Omega\|_{L^p(\Omega)} + \|u_\Omega - u_\varphi\|_{L^p(\Omega)}$$

and

$$u_\Omega - u_\varphi = \int_\Omega u \left(\frac{1}{|\Omega|} - \varphi \right) = \int_\Omega (u - u_\Omega) \left(\frac{1}{|\Omega|} - \varphi \right),$$

where in the last equality we used that $\int_\Omega \frac{1}{\Omega} - \varphi = 0$. Therefore, using the Hölder inequality (recall that φ is bounded) we see that $\|u - u_\varphi\|_{L^p(\Omega)} \leq C \|u - u_\Omega\|_{L^p(\Omega)}$. Finally we show that $\|u - u_\Omega\|_{L^p(\Omega)} \leq 2\|u - c\|_{L^p(\Omega)}$ for a generic c. To this end we write

$$\|u - u_\Omega\|_{L^p(\Omega)} \leq \|u - c\|_{L^p(\Omega)} + \|c - u_\Omega\|_{L^p(\Omega)}.$$

The first term is fine, for the second observe that $c - u_\Omega = \frac{1}{\Omega} \int_\Omega (c - u)$ and hence $\|c - u_\Omega\|_{L^p(\Omega)} \leq \|c - u\|_{L^p(\Omega)}$. Therefore $\|u - u_\Omega\|_{L^p(\Omega)} \leq 2\|u\|_{[L^p(\Omega)]}$ and (1) follows.

(2) We just show that for any weight ω,

$$\|u - u_\Omega\|_{L^p(\Omega, \omega)} \leq C \|u - u_\omega\|_{L^p(\Omega, \omega)}.$$

Using the triangle inequality we see that it is enough to bound $\|u_\Omega - u_\omega\|_{L^p(\omega)}$ then

$$\|u_\Omega - u_\omega\|_{L^p(\Omega,\omega)} = \omega(\Omega)^{1/p}|u_\Omega - u_\omega| = \omega(\Omega)^{1/p}\left|\int_\Omega u\left(\frac{1}{|\Omega|} - \frac{\omega}{\omega(\Omega)}\right)\right|$$

therefore

$$\|u_\Omega - u_\omega\|_{L^p(\Omega,\omega)} = \omega(\Omega)^{1/p}\left|\int_\Omega (u - u_\omega)\left(\frac{1}{|\Omega|} - \frac{\omega}{\omega(\Omega)}\right)\right|.$$

Multiplying and dividing by $\omega^{1/p}$ inside the last integral and a further use of Hölder inequality show that it is enough to check $\int_\Omega \left(\frac{1}{|\Omega|} - \frac{\omega}{\omega(\Omega)}\right)^q \omega^{-\frac{q}{p}} \leq C$, but this follows easily from the integrability of ω and $\omega^{-\frac{1}{p-1}}$. $\square$

Remark 3.2. Notice that thanks to Lemma 3.1 a Poincaré domain can be defined using any of the expressions $\|u\|_{[L^p(\Omega,\omega_1)]}, \|u - u_\varphi\|_{L^p(\Omega,\omega_1)}, \|u - u_\Omega\|_{L^p(\Omega,\omega_1)}, \|u - u_{\Omega,\omega_1}\|_{L^p(\Omega,\omega_1)}, \|u - u_{\varphi\omega_1}\|_{L^p(\Omega,\omega_1)}$ on the left-hand side.

Remark 3.3. A_p weights (2.2.30) verify the conditions required in the previous lemma.

Observe that $Korn_p$ is equivalent to asking for (3.1.2) for any $\mathbf{w} \in [\mathscr{S}]$. Similarly, it is easy to see that the same holds for $Korn_p^u$ and (3.1.3). That is, $Korn_p^u$ holds if and only if

$$\|\mathbf{Dw}\|_{L^p(\Omega)^n} \leq C\left\{\|\mathbf{w}\|_{[L^p(\Omega)^n]} + \|\boldsymbol{\varepsilon}(\mathbf{w})\|_{L^p(\Omega)^n}\right\}, \tag{3.1.4}$$

for any $\mathbf{w} \in [W^{1,p}(\Omega)]$.

At this point we want to mention the following decomposition

$$[\mathscr{S}] \oplus [RM(\Omega)] = [W^{1,p}(\Omega)^n], \tag{3.1.5}$$

which can be proved as follows. Given $\mathbf{u} \in W^{1,p}(\Omega)^n$, we can pick $\mathbf{v} \in RM$ as

$$\mathbf{v} = \mathbf{Mx} \tag{3.1.6}$$

being the matrix $\mathbf{M} = \boldsymbol{\mu}(\mathbf{u})_\Omega$. Obviously $\mathbf{w} = \mathbf{u} - \mathbf{v} \in \mathscr{S}$ and for any other element $\tilde{\mathbf{v}} \in RM$ such that $\mathbf{u} - \tilde{\mathbf{v}} \in \mathscr{S}$ we have, due to the kind of elements belonging to RM, that $\mathbf{v} - \tilde{\mathbf{v}}$ is constant (since $\boldsymbol{\mu}(\mathbf{v}) = \boldsymbol{\mu}(\tilde{\mathbf{v}})$). As a consequence, $\mathbf{v}$ can be uniquely defined up to an additive constant. Taking into account that for any constant $\mathbf{c}$, $\mathbf{u} + \mathbf{c}$ leads to the same choice of $\mathbf{v}$ we have by construction

$$\|\mathbf{v}\|_{[W^{1,p}(\Omega)^n]} \leq C\|\mathbf{u}\|_{[W^{1,p}(\Omega)^n]} \qquad \|\mathbf{w}\|_{[W^{1,p}(\Omega)^n]} \leq C\|\mathbf{u}\|_{[W^{1,p}(\Omega)^n]}. \tag{3.1.7}$$

Decomposition (3.1.5) and inequalities (3.1.7) are useful in the sequel. In the meantime let us introduce an inequality stronger than (3.1.3).

Let B be a ball $B \subset \Omega$, if the following inequality

$$\|\mathbf{Du}\|_{L^p(\Omega)^n} \le C \Big\{ \|\mathbf{u}\|_{L^p(B)^n} + \|\boldsymbol{\varepsilon}(\mathbf{u})\|_{L^p(\Omega)^n} \Big\}, \tag{3.1.8}$$

holds for any $\mathbf{u} \in W^{1,p}(\Omega)^n$ we say that Ω satisfies $Korn^u_{p,B}$ or just that $Korn^u_{p,B}$ holds. Now we turn our attention to the second Korn's inequality. Let us first recall the quotient space $[L^p(\Omega)^{n \times n}]_{Skew} := L^p(\Omega)^{n \times n}/Skew$ with the norm

$$\|\mathbf{M}(x)\|_{[L^p(\Omega)^{n \times n}]_{Skew}} := \inf_{\boldsymbol{sym}(\mathbf{R})=0} \|\mathbf{M}(x) - \mathbf{R}\|_{L^p(\Omega)^{n \times n}}.$$

Lemma 3.2. *With the same notation of Lemma 3.1, taking into account the obvious adaptation to matrices, we have for any* $\mathbf{M}(x) \in L^p(\Omega)^{n \times n}$

$$\|\mathbf{M}(x)\|_{[L^p(\Omega)^{n \times n}]_{Skew}} \sim \|\mathbf{M}(x) - \boldsymbol{\mu}(\mathbf{M})_\varphi\|_{L^p(\Omega)^{n \times n}} \sim \|\mathbf{M}(x) - \boldsymbol{\mu}(\mathbf{M})_\Omega\|_{L^p(\Omega)^{n \times n}}.$$

Proof. The proof is essentially that of Lemma 3.1, just recalling that $\mathbf{R} = \boldsymbol{skew}(\mathbf{R})$ for any matrix $\mathbf{R} \in \mathbb{R}^{n \times n}$ such that $\boldsymbol{sym}(\mathbf{R}) = 0$. □

Notice that, thanks to Lemma 3.2, $Korn_p$ can be stated as

$$\|\mathbf{Du}\|_{[L^p(\Omega)^{n \times n}]_{Skew}} \le C \|\boldsymbol{\varepsilon}(\mathbf{u})\|_{L^p(\Omega)^{n \times n}}. \tag{3.1.9}$$

There is a standard argument showing that $Korn_p$ implies $Korn^u_{p,B}$. It assumes that Ω is a Poincaré domain. The argument is essentially the following: if $Korn^u_{p,B}$ does not hold, we can find a sequence $\{\mathbf{u}_n\} \subset [W^{1,p}(\Omega)^n]$ such that

$$\|\mathbf{Du}_n\|_{L^p(\Omega)^{n \times n}} = 1 \tag{3.1.10}$$

while

$$\|\mathbf{u}_n\|_{[L^p(B)^n]} + \|\boldsymbol{\varepsilon}(\mathbf{u}_n)\|_{L^p(\Omega)^{n \times n}} < \frac{1}{n}. \tag{3.1.11}$$

We show now that there exists a subsequence of $\mathbf{u}_n$ that converges strongly in $[W^{1,p}(\Omega)]$ to some $\mathbf{v} \in [W^{1,p}(\Omega)^n]$.

From (3.1.5) we may write

$$\mathbf{u}_n = \mathbf{v}_n + \mathbf{w}_n$$

with $\mathbf{v}_n \in [RM]$ and $\mathbf{w}_n \in [\mathscr{S}]$. Now, from Poincaré inequality, the fact that (3.1.2) is valid in $[\mathscr{S}]$ and equation (3.1.11) we get

$$\|\mathbf{w}_n\|_{[W^{1,p}(\Omega)^n]} \le C \|\mathbf{Dw}_n\|_{L^p(\Omega)^{n \times n}} \le C \|\boldsymbol{\varepsilon}(\mathbf{w}_n)\|_{L^p(\Omega)^{n \times n}} < C \frac{1}{n}, \tag{3.1.12}$$

since $\boldsymbol{\varepsilon}(\mathbf{u}_n) = \boldsymbol{\varepsilon}(\mathbf{w}_n)$ and therefore $\mathbf{w}_n \to 0$ in $[W^{1,p}(\Omega)^n]$. On the other hand, thanks to (3.1.7), $\mathbf{v}_n$ is bounded in $[W^{1,p}(\Omega)^n]$ and belongs to the *finite dimensional space* $[RM]$. Therefore there exists a subsequence, called again $\mathbf{v}_n$, such that $\mathbf{v}_n \longrightarrow \mathbf{v}$ strongly in $[W^{1,p}(\Omega)^n]$. As a consequence $\mathbf{u}_n$ converges strongly in $[W^{1,p}(\Omega)^n]$ to

v and taking into account that $\|\cdot\|_{[L^p(B)^n]}$ defines an equivalent norm to $\|\cdot\|_{[L^p(\Omega)^n]}$ in the finite dimensional space RM, we get from (3.1.11) that $\|\mathbf{v}\|_{[L^p(\Omega)^n]} = 0$. while from (3.1.10) $\|\mathbf{Dv}\|_{L^p(\Omega)^n} = 1$, a contradiction.

Remark 3.4. Recall that the decomposition of a Banach space U into complementary closed subspaces $V, W \subset U$ guarantees the continuity of the projectors $u \to v$ and $u \to w$, where $u = v + w$ (see Section 2.4 in [15]). In the proof above this fact is given by (3.1.7). A careful reading of previous arguments shows the following: Let $\mathscr{S}$ be a closed subspace of $W^{1,p}(\Omega)^n$ containing the set of constant vector fields and such that

$$[\mathscr{S}] \oplus [RM] = [W^{1,p}(\Omega)^n]. \tag{3.1.13}$$

If Ω is a *Poincaré* domain and $Korn_p$ holds in $\mathscr{S}$, then $Korn_p^u$ holds.

Previous remark applies for an abstract $\mathscr{S}$. Theorem 3.1 below shows that for a *classical* second case a slightly different approach avoids the use of Poincaré inequality. As a consequence this hypothesis is not needed in its statement.

Theorem 3.1. $Korn_p$ *implies* $Korn_{p,B}^u$.

Proof. Let φ be a smooth function supported in B and such that $\int_\Omega \varphi = 1$. For any $\mathbf{u} \in W^{1,p}(\Omega)^n$ we consider $\mathbf{v}$ given by (3.1.6) with $\mathbf{M} = \int_\Omega \boldsymbol{\mu}(\mathbf{u})\varphi = \boldsymbol{\mu}(\mathbf{u})_\varphi$. Define $\mathbf{w} = \mathbf{u} - \mathbf{v}$ and notice that $\boldsymbol{\varepsilon}(\mathbf{w}) = \boldsymbol{\varepsilon}(\mathbf{u})$. Using Lemma 3.2 we see that $Korn_p$ implies

$$\|\mathbf{Dw}\|_{L^p(\Omega)^{n\times n}} \leq C\|\boldsymbol{\varepsilon}(\mathbf{w})\|_{L^p(\Omega)^{n\times n}} = C\|\boldsymbol{\varepsilon}(\mathbf{u})\|_{L^p(\Omega)^{n\times n}}.$$

On the other hand, since φ *is supported in* B, we get

$$\|\mathbf{Dv}\|_{L^p(\Omega)^{n\times n}} \leq C\|\boldsymbol{\mu}(\mathbf{u})_\varphi\|_\infty |\Omega|^{1/p} \leq C\|\mathbf{u}\|_{L^p(B)^n}$$

where $\|\cdot\|_\infty$ is a matrix norm and integration by parts allows to bound each element of $\boldsymbol{\mu}(\mathbf{u})_\varphi$. Therefore $Korn_{p,B}^u$ follows by means of the triangle inequality. $\square$

Since $Korn_{p,B}^u$ implies $Korn_p^u$, we have

Corollary 3.1. $Korn_p$ *implies* $Korn_p^u$.

Well-known arguments give the implication $Korn_p^u \implies Korn_p$ in domains for which the compact inclusion $L^p(\Omega) \hookrightarrow W^{1,p}(\Omega)$ holds. Indeed, suppose that (3.1.2) fails in the second case. Then, there exists a sequence $\{\mathbf{u}_n\} \subset [\mathscr{S}]$ such that each $\mathbf{u}_n$ verifies (3.1.10) while

$$\|\boldsymbol{\varepsilon}(\mathbf{u}_n)\|_{L^p(\Omega)^{n\times n}} < \frac{1}{n}. \tag{3.1.14}$$

Moreover, from the embedding $L^p(\Omega) \hookrightarrow W^{1,p}(\Omega)$ we know that Ω is a Poincaré domain and therefore $\mathbf{u}_n$ is bounded in $[W^{1,p}(\Omega)^n]$. As a consequence there exists a subsequence, called again $\mathbf{u}_n$, that converges weakly in $[W^{1,p}(\Omega)^n]$ and strongly in $[L^p(\Omega)]$ to some $\mathbf{v} \in [W^{1,p}(\Omega)^n]$. In particular we conclude that $\boldsymbol{\varepsilon}(\mathbf{v}) = 0$ in Ω, which in turn says that $\mathbf{v} \in RM$. Applying now (3.1.3) to $\mathbf{u}_n - \mathbf{v}$ we get that actually $\mathbf{u}_n \to \mathbf{v}$ strongly in $[W^{1,p}(\Omega)^n]$, and therefore $\mathbf{v} \in [\mathscr{S}] \cap [RM] = \{0\}$ while from (3.1.10) $\|\mathbf{Dv}\|_{L^p(\Omega)^{n\times n}} = 1$, a contradiction. Then we have proved.

Theorem 3.2. *Assume the compact inclusion $L^p(\Omega) \hookrightarrow W^{1,p}(\Omega)$. Then $Korn_p^u$ implies $Korn_p$.*

Remark 3.5. From the proof of Theorem 3.2 we notice that the following holds: Let Ω be a domain for which the compact embedding $L^p(\Omega) \hookrightarrow W^{1,p}(\Omega)$ holds. Let $\mathscr{S}$ be a closed subspace of $W^{1,p}(\Omega)^n$ containing the set of constant vector fields.
If
$$[\mathscr{S}] \cap [RM(\Omega)] = \{0\},$$
then $Korn_{p,B}^u$ implies $Korn_p$ for any $\mathbf{u} \in \mathscr{S}$.

Theorem 3.3 below is the reciprocal of Theorem 3.1. Notice that the compact embedding $L^p(\Omega) \hookrightarrow W^{1,p}(\Omega)$ is not necessary.

Theorem 3.3. *$Korn_{p,B}^u$ implies $Korn_p$.*

Proof. It is enough to show (3.1.9). To this end we pick a smooth function φ supported in B and such that $\int_\Omega \varphi = 1$. Take any $\mathbf{u} \in W^{1,p}(\Omega)$, define $\mathbf{M} = Skew(\mathbf{Du})_\varphi$ and consider $\mathbf{v} = \mathbf{M}(\mathbf{x} - \mathbf{x}_c) + \mathbf{p}$ with $\mathbf{x}_c$ the barycenter of Ω and $\mathbf{p} = \mathbf{u}_\varphi$. Applying $Korn_{p,B}^u$ to $\mathbf{w} := \mathbf{u} - \mathbf{v}$ we get

$$\|\mathbf{Du}\|_{[L^p(\Omega)^{n\times n}]_{Skew}} \le \|\mathbf{Dw}\|_{L^p(\Omega)^{n\times n}} \le C\|\boldsymbol{\varepsilon}(\mathbf{u})\|_{L^p(\Omega)^{n\times n}} + \|\mathbf{w}\|_{L^p(B)^n},$$

where we have used that $\boldsymbol{\varepsilon}\mathbf{w} = \boldsymbol{\varepsilon}\mathbf{u}$. On the other hand, $\mathbf{w}$ verifies both Poincaré and second case of Korn *in the ball B* and hence

$$\|\mathbf{w}\|_{L^p(B)^n} \le C\|\mathbf{Dw}\|_{L^p(B)^{n\times n}} \le C\|\boldsymbol{\varepsilon}(\mathbf{u})\|_{L^p(B)^{n\times n}}.$$

Using this bound in previous inequality gives the desired result. □

Remark 3.6. In [58] it is shown that if $Korn_{p,B}^u$ holds for a ball $B \subset\subset \Omega$, then Ω is a Poincaré domain.

Finally we briefly discuss another case of the Korn inequality arising in the context of mixed boundary conditions for the linearized elasticity equations.

For $\Gamma \subset \partial\Omega$, with positive $n-1$ Hausdorff measure, and such that the trace operator $\mathscr{T}: W^{1,p}(\Omega) \to L^p(\Gamma)$ is well defined, we consider the space $\mathscr{S}_\Gamma := W_\Gamma^{1,p}(\Omega)^n$. We say that Ω verifies $Korn_{p,\Gamma}$ or that $Korn_{p,\Gamma}$ holds in Ω if for any $\mathbf{u} \in \mathscr{S}_\Gamma$ equation (3.1.2) holds. As before, the reference to Ω may be omitted if it is clear from the context. Obviously, the *first case* of Korn's inequality can be regarded as $Korn_{p,\Gamma}$ with $\Gamma = \partial\Omega$, although in that case no regularity on the boundary is required. In the remaining part of this section we focus on the case in which Γ is a proper subset of the boundary.

Notice that $\mathscr{S}_\Gamma$ *does not* contain the set of constant vector fields. We have the following.

Lemma 3.3. $RM \cap \mathscr{S}_\Gamma = \{0\}$

Proof. Let $\mathbf{v} \in RM \cap \mathscr{S}_\Gamma$, that is $\mathbf{v} = \mathbf{M}\mathbf{x} + \mathbf{p}$ with $\mathbf{M}^t = -\mathbf{M}$ while $\mathbf{v} = 0$ on Γ. Therefore $\Gamma \subset \mathbb{H}$ with $\mathbb{H} := \{\mathbf{x} \in \mathbb{R}^n : \mathbf{M}\mathbf{x} = -\mathbf{p}\}$. Since the $n-1$ Hausdorff measure of Γ is positive then we have that $\mathbb{H}$ is either a hyperplane or the whole $\mathbb{R}^n$. We claim that $\mathbb{H} = \mathbb{R}^n$, which in turn says that $\mathbf{v} = 0$. First observe that since $\mathbf{M}^t = -\mathbf{M}$ then nontrivial eigenvalues of $\mathbf{M}$ come in pairs of the form $\xi, -\xi \in \mathbb{C}$ due to the fact that the characteristic polynomials of $\mathbf{M}$ and $\mathbf{M}^t$ are the same. In order to finish our arguments we find easier to work with the *Hermitian* matrix $i\mathbf{M}$. Indeed, on the one hand the eigenvalues $i\mathbf{M}$ are real and on the other hand those which are nontrivial should come in pairs $\lambda, -\lambda$ (since they agree with $i\xi$ where ξ is eigenvalue of $\mathbf{M}$). As $i\mathbf{M}$ can be diagonalized we notice that its rank agrees with the number of eigenvalues different of zero and as a consequence the rank of $i\mathbf{M}$ is even. Therefore, the parity of the dimension of the kernel of $i\mathbf{M}$ -and hence of $\mathbf{M}$- is the same as that of $\mathbb{R}^n$. We conclude that $\mathbb{H}$ cannot be a hyperplane and the lemma follows.   □

As a consequence, with the same proof of Theorem 3.2 we have the following result.

Theorem 3.4. *Assume the compact inclusion $L^p(\Omega) \hookrightarrow W^{1,p}(\Omega)$, then $Korn_p^u$ implies $Korn_{p,\Gamma}$.*

To summarize what we have proved so far:

$$\text{Korn}_p \iff \text{Korn}_{p,B}^u \implies \text{Korn}_p^u$$

holds under no extra hypotheses.

On the other hand, if the compact embedding $L^p(\Omega) \hookrightarrow W^{1,p}(\Omega)$ holds, then

$$\text{Korn}_p^u \implies \text{Korn}_p,$$

and in particular all of them are equivalent.

If in addition we have a set $\Gamma \subset \partial\Omega$, with positive $n-1$ Hausdorff measure, for which the trace operator $\mathscr{T} : W^{1,p}(\Omega) \to L^p(\Gamma)$ is well defined, then it also holds

$$\text{Korn}_p^u \implies \text{Korn}_{p,\Gamma}.$$

3.2 The Korn Inequality on John Domains

In this section we use the results for the divergence operator given in Chapter 2 to show that Korn's inequality holds on John domains. This is done below, although two alternative proofs are also provided along this section.

Theorem 3.5. *Let $1 < p < \infty$. If div_p holds in Ω, then $Korn_p$, $Korn_{p,B}^u$, and $Korn_p^u$ also hold in Ω. In particular, they are valid on John domains.*

Proof (First proof). Thanks to Theorem 3.1 and Corollary 3.1 we only need to concentrate on the second case of Korn's inequality.

Let $\mathbf{v} \in W^{1,p}(\Omega)^n$ such that (3.1.1) holds. By density we can assume that $\mathbf{v}$ is smooth. We have

$$\|\mathbf{Dv}\|_{L^p(\Omega)^{n\times n}} \leq \|\boldsymbol{\varepsilon}(\mathbf{v})\|_{L^p(\Omega)^{n\times n}} + \|\boldsymbol{\mu}(\mathbf{v})\|_{L^p(\Omega)^{n\times n}}$$

and so it is enough to prove that

$$\|\boldsymbol{\mu}(\mathbf{v})\|_{L^p(\Omega)^{n\times n}} \leq C\|\boldsymbol{\varepsilon}(\mathbf{v})\|_{L^p(\Omega)^{n\times n}}. \tag{3.2.1}$$

In order to show this, for any pair $1 \leq i, j \leq n$, we write by duality

$$\|\boldsymbol{\mu}_{ij}(\mathbf{v})\|_{L^p(\Omega)} = \int_\Omega \boldsymbol{\mu}_{ij}(\mathbf{v})\tilde{\boldsymbol{\mu}}_{ij}, \tag{3.2.2}$$

for an appropriate $\tilde{\boldsymbol{\mu}}_{ij} \in L^{p'}(\Omega)$, $\|\tilde{\boldsymbol{\mu}}_{ij}\|_{L^{p'}(\Omega)} = 1$. Moreover, thanks to (3.1.1), we may assume that $\tilde{\boldsymbol{\mu}}_{i,j} \in L_0^{p'}(\Omega)$ and therefore it is possible to solve the problem

$$\operatorname{div} \mathbf{u}^{ij} = \tilde{\boldsymbol{\mu}}_{ij}$$

with $\mathbf{u}^{ij} \in W_0^{1,p'}(\Omega)$ for which

$$\|\mathbf{Du}^{ij}\|_{L^{p'}(\Omega)} \leq C\|\tilde{\boldsymbol{\mu}}_{ij}\|_{L^{p'}(\Omega)}. \tag{3.2.3}$$

Then,

$$\|\boldsymbol{\mu}_{ij}(\mathbf{v})\|_{L^p(\Omega)} = \int_\Omega \boldsymbol{\mu}_{ij}(\mathbf{v})\operatorname{div}\mathbf{u}^{ij} = -\int_\Omega \nabla\boldsymbol{\mu}_{ij}(\mathbf{v})\cdot\mathbf{u}^{ij}$$

but,

$$\frac{\partial\boldsymbol{\mu}_{ij}(\mathbf{v})}{\partial x_k} = \left(\frac{\partial\boldsymbol{\varepsilon}_{ik}(\mathbf{v})}{\partial x_j} - \frac{\partial\boldsymbol{\varepsilon}_{jk}(\mathbf{v})}{\partial x_i}\right)$$

and taking into account,

$$-\sum_{k=1}^{n}\int_\Omega\left(\frac{\partial\boldsymbol{\varepsilon}_{ik}(\mathbf{v})}{\partial x_j} - \frac{\partial\boldsymbol{\varepsilon}_{jk}(\mathbf{v})}{\partial x_i}\right)u_k^{ij} = \sum_{k=1}^{n}\int_\Omega\left(\boldsymbol{\varepsilon}_{ik}(\mathbf{v})\frac{\partial u_k^{ij}}{\partial x_j} - \boldsymbol{\varepsilon}_{jk}(\mathbf{v})\frac{\partial u_k^{ij}}{\partial x_i}\right),$$

together with (3.2.3) and the fact that $\|\tilde{\boldsymbol{\mu}}_{ij}\|_{L^{p'}(\Omega)} = 1$ we obtain

$$\|\boldsymbol{\mu}_{ij}(\mathbf{v})\|_{L^p(\Omega)} \leq C\sum_{k=1}^{n}\left(\|\boldsymbol{\varepsilon}_{ik}(\mathbf{v})\|_{L^p(\Omega)} + \|\boldsymbol{\varepsilon}_{jk}(\mathbf{v})\|_{L^p(\Omega)}\right).$$

Summing in i and j we obtain (3.2.1).   □

Another proof for the previous theorem can be written directly using the L^p version of item (5) in Proposition 1.1. For the sake of completeness and further use we give the following lemma in weighted spaces. Observe that in (3.2.4) the negative norm of f is restricted to a ball.

Lemma 3.4. *Let ω be a weight such that ω and $\omega^{1-p'}$ are integrable in a bounded domain $\Omega \subset \mathbb{R}^n$. Assume that for any $1 < p < \infty$ and any $g \in L_0^{p'}(\Omega, \omega^{1-p'})$ there exists $\mathbf{u} \in W_0^{1,p'}(\Omega, \omega^{1-p'})^n$ such that $\operatorname{div} \mathbf{u} = g$ and*

$$\|\mathbf{u}\|_{W^{1,p'}(\Omega, \omega^{1-p'})^n} \leq C\|g\|_{L^{p'}(\Omega, \omega^{1-p'})},$$

with a constant C depending only on Ω, p, and ω. Fix an open ball $B \subset \Omega$. Then, for any $f \in L^p(\Omega, \omega)$,

$$\|f\|_{L^p(\Omega, \omega)} \leq C\left\{\|f\|_{W^{-1,p}(B)} + \|\nabla f\|_{W^{-1,p}(\Omega, \omega)^n}\right\}, \tag{3.2.4}$$

where the constant C depends only on Ω, B, p, and ω.

Proof. The proof follows arguments used in Chapter 1: take $f \in L^p(\Omega, \omega)$, we have for $g \in L^{p'}(\Omega, \omega^{1-p'})$,

$$\int_\Omega (f - f_\Omega)g = \int_\Omega (f - f_\Omega)(g - g_\Omega).$$

But, from our hypothesis, there exists a solution $\mathbf{u} \in W_0^{1,p'}(\Omega, \omega^{1-p'})^n$ of $\operatorname{div} \mathbf{u} = g - g_\Omega$ satisfying

$$\|\mathbf{u}\|_{W^{1,p'}(\Omega, \omega^{1-p'})^n} \leq C\|g - g_\Omega\|_{L^{p'}(\Omega, \omega^{1-p'})}.$$

Thus,

$$\int_\Omega (f - f_\Omega)g = \int_\Omega (f - f_\Omega)\operatorname{div}\mathbf{u} \leq \|\nabla f\|_{W^{-1,p}(\Omega, \omega)^n}\|\mathbf{u}\|_{W^{1,p'}(\Omega, \omega^{1-p'})^n}$$
$$\leq C\|\nabla f\|_{W^{-1,p}(\Omega, \omega)^n}\|g - g_\Omega\|_{L^{p'}(\Omega, \omega^{1-p'})}.$$

Therefore, by duality,

$$\|f - f_\Omega\|_{L^p(\Omega, \omega)} \leq C\|\nabla f\|_{W^{-1,p}(\Omega, \omega)^n}. \tag{3.2.5}$$

Now, we decompose f as

$$f = (f - f_\varphi) + f_\varphi,$$

with $\varphi \in C_0^\infty(B)$ such that $\int_B \varphi = 1$ and $f_\varphi := \int_B f\varphi$. Thanks to Lemma 3.1 and equation (3.2.5) we have

$$\|f - f_\varphi\|_{L^p(\Omega, \omega)} \leq C\|\nabla f\|_{W^{-1,p}(\Omega, \omega)^n}.$$

On the other hand,

$$\|f_\varphi\|_{L^p(\Omega, \omega)} \leq \omega(\Omega)^{1/p}\left|\int_B f\varphi\right| \leq \omega(\Omega)^{1/p}\|f\|_{W^{-1,p}(B)}\|\varphi\|_{W_0^{1,p'}(B)},$$

where $\omega(\Omega)$ stands for $\int_\Omega \omega$. As a consequence, the lemma follows. $\qquad\square$

An immediate consequence is the following.

Proof (Second proof of Theorem 3.5:). $(Korn_{p,B}^u$ from (3.2.4))

Using (3.2.4) with $\omega = 1$ we get for any $1 \leq i \leq n$,

$$\left\| \frac{\partial u_i}{\partial x_j} \right\|_{L^p(\Omega)} \leq C \left(\left\| \frac{\partial u_i}{\partial x_j} \right\|_{W^{-1,p}(B)} + \left\| \nabla \frac{\partial u_i}{\partial x_j} \right\|_{W^{-1,p}(\Omega)^n} \right),$$

therefore, using (1.0.6),

$$\left\| \frac{\partial u_i}{\partial x_j} \right\|_{L^p(\Omega)} \leq C \left(\|u_i\|_{L^p(B)} + \sum_{1 \leq k \leq l \leq n} \|\nabla \varepsilon_{kl}(\mathbf{u})\|_{W^{-1,p}(\Omega)^n} \right).$$

Then, $Korn_{p,B}^u$ follows immediately.   □

Yet another interesting approach can be used to deal with Korn's inequalities on John domains. It involves improved Poincaré inequalities in a fashion that can be traced back to [65]. This approach also provides weighted versions of Korn in Hölder domains [1], a topic that is addressed in the next chapter. In what follows we will make use of Theorem 2.9. We need also the following result for harmonic functions borrowed from [31] ([65] for a different proof in the case $p = 2$ and $\mu = 0$).

Lemma 3.5. *Let $\Omega \subset \mathbb{R}^n$ be an arbitrary bounded domain and $1 \leq p < \infty$. If f is a harmonic function in Ω, then*

$$\|d^{1-\mu}\nabla f\|_{L^p(\Omega)} \leq C\|d^{-\mu}f\|_{L^p(\Omega)}$$

for all $\mu \in \mathbb{R}$.

Proof. Given $x \in \Omega$, let $B(x,R) \subset \Omega$ be the ball with center at x and radius R. Since f is harmonic in Ω it satisfies the following inequality (see, for example, [39]),

$$|\nabla f(x)|^p \leq \frac{C}{R^{n+p}} \int_{B(x,R)} |f(y)|^p dy$$

Let us take $R = d(x)/2$ in this inequality. Then we have

$$\int_\Omega |\nabla f(x)|^p d(x)^{p(1-\mu)} dx \leq C \int_\Omega d(x)^{-n-p\mu} \left(\int_{B(x,d(x)/2)} |f(y)|^p dy \right) dx$$

But, since $|d(x) - d(y)| \leq |x - y|$, we have that $\frac{d(x)}{2} \leq d(y) \leq \frac{3}{2}d(x)$ whenever $|x - y| < \frac{d(x)}{2}$. Therefore, we can change the order of integration and replace $d(x)$ by $d(y)$ to obtain

$$\int_\Omega |\nabla f(x)|^p d(x)^{p(1-\mu)} dx \leq C \int_\Omega |f(y)|^p d(y)^{-n-p\mu} \left(\int_{B(y,d(y))} dx \right) dy$$

hence

$$\int_\Omega |\nabla f(x)|^p d(x)^{p(1-\mu)} dx \leq C \int_\Omega |f(y)|^p d(y)^{-p\mu} dy,$$

which is the required inequality.   □

Proof (Third proof of Theorem 3.5). ($Korn^u_{p,B}$ from Improved Poincaré) Following [65] we can show that there exists $\mathbf{v} \in W^{1,p}(\Omega)^n$ such that

$$\Delta \mathbf{v} = \Delta \mathbf{u} \qquad \text{in } \Omega \tag{3.2.6}$$

and

$$\|\mathbf{v}\|_{W^{1,p}(\Omega)^n} \leq C \|\boldsymbol{\varepsilon}(\mathbf{u})\|_{L^p(\Omega)^{n \times n}} \tag{3.2.7}$$

Indeed, define

$$F = \begin{cases} 2\boldsymbol{\varepsilon}(\mathbf{u}) - (\operatorname{tr}\boldsymbol{\varepsilon}(\mathbf{u}))I & \text{in } \Omega \\ 0 & \text{outside } \Omega, \end{cases}$$

Then, it is easy to check that $\operatorname{Div} F = \Delta \mathbf{u}$ in Ω and so, one can obtain $\mathbf{v}$ by solving a Poisson equation in a smooth domain, for example a ball B_1, containing Ω. In fact, since $\operatorname{Div} F \in W^{-1,p}(B_1)^n$, there exists $\mathbf{v} \in W_0^{1,p}(B_1)^n$ such that

$$\Delta \mathbf{v} = \operatorname{Div} F$$

and (3.2.7) is satisfied in view of known a priori estimates for smooth domains.

Now, let B be a ball contained in Ω and $\varphi \in C_0^\infty(B)$ be such that $\int_B \varphi \, dx = 1$. Define the linear mapping $\mathbf{L}(\mathbf{x}) := \mathbf{M}\mathbf{x}$ by means of the matrix $\mathbf{M} = \mathbf{D}(\mathbf{u} - \mathbf{v})_\varphi$

Then we have

$$\mathbf{M} = \mathbf{DL} = \int_B \mathbf{D}(\mathbf{u} - \mathbf{v}) \varphi \, dx$$

and so, integrating by parts, we obtain

$$\|\mathbf{M}\|_\infty \leq C \|\mathbf{u} - \mathbf{v}\|_{L^p(B)^n} \|\nabla \varphi\|_{L^{p'}(B)^n}$$

where p' is the dual exponent of p. Therefore, it follows from (3.2.7) that there exists a constant C depending only on Ω, p and φ such that

$$\|\mathbf{DL}\|_{L^p(\Omega)^{n \times n}} \leq C \left\{ \|\boldsymbol{\varepsilon}(\mathbf{u})\|_{L^p(\Omega)^{n \times n}} + \|\mathbf{u}\|_{L^p(B)^n} \right\} \tag{3.2.8}$$

Let us now introduce

$$\mathbf{w} = \mathbf{u} - \mathbf{v} - \mathbf{L}$$

Then, in view of the bounds (3.2.7) and (3.2.8), to conclude the proof we have to estimate $\mathbf{w}$. But, from (3.2.6) and the fact that $\mathbf{L}$ is linear we know that

$$\Delta \mathbf{w} = 0$$

and consequently

$$\Delta \boldsymbol{\varepsilon}(\mathbf{w}) = 0$$

therefore, we can apply Lemma 3.5 with $\mu = 0$ to obtain

$$\|d \nabla \boldsymbol{\varepsilon}_{ij}(\mathbf{w})\|_{L^p(\Omega)} \leq C \|\boldsymbol{\varepsilon}_{ij}(\mathbf{w})\|_{L^p(\Omega)} \tag{3.2.9}$$

and using the key identity (1.0.6), we conclude that for any $1 \leq i \leq n$

$$\|d\mathbf{D}(\nabla \mathbf{w}_i)\|_{L^p(\Omega)^{n\times n}} \leq C\|\boldsymbol{\varepsilon}(\mathbf{w})\|_{L^p(\Omega)^{n\times n}} \tag{3.2.10}$$

Since $\int_\Omega \mathbf{D}\mathbf{w}\varphi\,dx = 0$ (indeed, we have defined $\mathbf{L}$ in order to have this property), it follows from Theorem 2.9 that for any $1 \leq i \leq n$,

$$\|\nabla \mathbf{w}_i\|_{L^p(\Omega)^n} \leq C\|d\mathbf{D}\nabla \mathbf{w}_i\|_{L^p(\Omega)^{n\times n}} \tag{3.2.11}$$

and therefore

$$\|\mathbf{D}\mathbf{w}\|_{L^p(\Omega)^{n\times n}} \leq C\|\boldsymbol{\varepsilon}(\mathbf{w})\|_{L^p(\Omega)^{n\times n}}$$

which together with (3.2.7) and (3.2.8) concludes the proof. $\qquad\qquad\square$

Remark 3.7 (Korn$^u_{p,B}$ with A_p weights). Notice that, following the second proof of Theorem 3.5, an application of Lemma 3.4 in weighted spaces in conjunction with Theorem 2.8 says that

$$\|\mathbf{D}\mathbf{u}\|_{L^p(\Omega,\omega)^{n\times n}} \leq C(\|\mathbf{u}\|_{L^p(B)^n} + \|\boldsymbol{\varepsilon}(\mathbf{u})\|_{L^p(\Omega,\omega)^{n\times n}}),$$

for $\omega \in A_p$.

We finish this section with other result useful in the next chapter. Its proof is exactly the same as that of Lemma 3.4 and therefore is omitted.

Lemma 3.6. *Let ω be a weight defined in a bounded domain $\Omega \subset \mathbb{R}^n$. Assume that for any $1 < p < \infty$ and any $g \in L_0^{p'}(\Omega)$ there exists $\mathbf{u} \in W_0^{1,p'}(\Omega, \omega^{1-p'})^n$ such that $\mathrm{div}\,\mathbf{u} = g$ and*

$$\|\mathbf{u}\|_{W^{1,p'}(\Omega,\omega^{1-p'})^n} \leq C\|g\|_{L^{p'}(\Omega)},$$

with a constant C depending only on Ω, p, and ω. Fix an open ball $B \subset \Omega$. Then, for any $f \in L^p(\Omega)$,

$$\|f\|_{L^p(\Omega)} \leq C\left\{\|f\|_{W^{-1,p}(B)} + \|\nabla f\|_{W^{-1,p}(\Omega,\omega)^n}\right\}, \tag{3.2.12}$$

where the constant C depends only on Ω, B, p, and ω.

Chapter 4
Singular Domains

In view of the previous chapters it is a natural question whether similar results can be obtained for more general domains. With this goal, we consider generalizations of $Korn_p$ and div_p using weighted Sobolev norms.

A motivation to work with weighted norms is given by the following theorem which generalizes the existence and uniqueness result proved in Theorem 1.1.

To state this result we first introduce some notation. Given a weight $\omega \in L^1(\Omega)$ we define

$$\mathbf{W} = \{\mathbf{v} \in H_0^1(\Omega)^n : \operatorname{div} \mathbf{v} \in L^2(\Omega, \omega^{-1})\}$$

which is a Hilbert space with the norm

$$\|\mathbf{v}\|_{\mathbf{W}}^2 = \|\mathbf{v}\|_{H_0^1(\Omega)^n}^2 + \|\operatorname{div} \mathbf{v}\|_{L^2(\Omega, \omega^{-1})}^2.$$

Also we will use the space $L_{\omega,0}^2(\Omega) = \{q \in L^2(\Omega, \omega) : \int_\Omega q\omega = 0\}$. Notice that in general $L_{\omega,0}^2(\Omega)$ and $L_0^2(\Omega, \omega)$ define different spaces.

Theorem 4.1. *Let $\Omega \subset \mathbb{R}^n$ be an arbitrary bounded domain and ω a weight which is integrable in Ω. There exists a constant C_1 depending only on Ω and ω, such that, for any $f \in L_0^2(\Omega)$,*

$$\|f\|_{[L^2(\Omega,\omega)]} \le C_1 \|\nabla f\|_{\mathbf{W}'}, \tag{4.0.1}$$

if and only if, for any $\mathbf{g} \in \mathbf{W}'$, there exists a unique solution $(\mathbf{u}, p) \in H_0^1(\Omega)^n \times L_{\omega,0}^2(\Omega)$ of the Stokes equations (1.0.14), and a constant C_2, depending only on Ω, ω, and μ, such that

$$\|\mathbf{u}\|_{H_0^1(\Omega)^n} + \|p\|_{L^2(\Omega,\omega)} \le C_2 \|\mathbf{g}\|_{\mathbf{W}'}$$

© The Author(s) 2017
G. Acosta, R.G. Durán, *Divergence Operator and Related Inequalities*,
SpringerBriefs in Mathematics, DOI 10.1007/978-1-4939-6985-2_4

Proof. The same arguments used to prove Theorem 1.1 can be applied here. Indeed, we have

$$|\langle \nabla q, \mathbf{v} \rangle| = \left| \int_{\Omega} q \operatorname{div} \mathbf{v} \right| \leq \|q\|_{L^2(\Omega,\omega)} \|\operatorname{div} \mathbf{v}\|_{L^2(\Omega,\omega^{-1})},$$

and therefore, ∇ defines a bounded operator from $[L^2(\Omega,\omega)]$ to $\mathbf{W}'$. Moreover, (4.0.1) implies that $Im\nabla$ is a closed subspace of $\mathbf{W}'$. Now the proof proceeds exactly as in Theorem 1.1. □

As in the unweighted case we will prove a result which is a dual version of (4.0.1).

Corollary 4.1. *Let* $\omega \in L^1(\Omega)$ *be a weight. Assume that for any* $f \in L^2_0(\Omega,\omega^{-1})$ *there exists* $\mathbf{u} \in H^1_0(\Omega)^n$ *such that div* $\mathbf{u} = f$ *and*

$$\|\mathbf{u}\|_{H^1(\Omega)^n} \leq C_1 \|f\|_{L^2(\Omega,\omega^{-1})},$$

with a constant C_1 *depending only on* Ω *and* ω. *Then, for any* $\mathbf{g} \in \mathbf{W}'$, *there exists a unique solution* $(\mathbf{u}, p) \in H^1_0(\Omega)^n \times L^2_{\omega,0}(\Omega)$ *of the Stokes equations (1.0.14). Moreover,*

$$\|\mathbf{u}\|_{H^1(\Omega)^n} + \|p\|_{L^2(\Omega,\omega)} \leq C_2 \|\mathbf{g}\|_{\mathbf{W}'},$$

where C_2 *depends only on* C_1 *and* Ω.

Proof. Given $f \in L^2(\Omega,\omega)$ such that $\int_{\Omega} f\omega = 0$ we solve div $\mathbf{v} = f\omega$ and then analogous arguments to those used in Lemma 3.4 implies (4.0.1). □

We begin by dealing with Hölder-α domains, looking for appropriate Korn's inequalities involving weights which are powers of the distance to the boundary of Ω. Then we show that if the domain is a cusp with one singular point, stronger inequalities with weights which are powers of the distance to the singularity can be obtained. Moreover, we consider *flat* cusps having a continuum of singular points (see Figure 4.1) and show how to adapt the previous results to this setting. To deal with these cases we prove a result that turns out to be important by its own: the existence of a right inverse of the divergence operator in weighted norms for this kind of domains. Next, we construct simple counterexamples showing that all our results are sharp. Finally, taking into account that div_p and $Korn_p$ hold on John domains, we consider generalized external cusps which are built by linking appropriate collections of John domains and show that the same kind of results obtained for standard external cusps can be obtained for these sets.

4.1 Weighted Korn and Poincaré Inequalities on Hölder α Domains

We start by noticing that the arguments used in the last section of the previous chapter can be extended to Hölder α domains. First we recall the following improved Poincaré inequality introduced in [12].

Lemma 4.1. *Let Ω be a bounded Hölder α domain, $0 < \alpha \le 1$. Then there exists a constant C depending only on Ω such that,*

$$\|f\|_{[L^p(\Omega)]} \le C\|\nabla f\|_{L^p(\Omega, d^{p\alpha})^n}. \tag{4.1.1}$$

Proof. See [12]. $\square$

In order to generalize Lemma 4.1 we exploit some ideas from [17]. Let us introduce for a set $A \subset \mathbb{R}^m$, $k \in \mathbb{N}$ and $0 \le t \le 1$

$$A^{k,t} = \{(\mathbf{x}, \mathbf{y}) \in A \times \mathbb{R}^k : |\mathbf{y}| < d_{\partial A}(\mathbf{x})^t\}.$$

In Appendix C we show that if A is a Hölder α domain the $A^{k,t}$ is also Hölder α. Taking this into account and following [1, Theorem 2.1] it is possible to extend Lemma 4.1.

First we prove the following result. Since several distance functions are involved we explicitly write $d_{\partial\Omega}$.

Lemma 4.2. *If $(x,y) \in \Omega^{k,t}$, then $d_{\partial\Omega^{k,t}}(x,y) \le d_\Omega(x)$.*

Proof. Obviously we can assume that the diameter of Ω is one. Given $y \in \mathbb{R}^k$, let $\Omega_y = \{(x,y) \in \mathbb{R}^n \times \mathbb{R}^k : x \in \Omega\}$. If $t = 0$ we have, for all $(x,y) \in \Omega^{k,0}$

$$d_{\partial\Omega}(x) = d_{\partial\Omega_y}(x,y) \ge d_{\partial\Omega^{k,0}}(x,y).$$

On the other hand, since the diameter of Ω is one, we have that for $0 \le t \le 1$,

$$\Omega^{k,t} \subset \Omega^{k,0}$$

and therefore, for $(x,y) \in \Omega^{k,t}$,

$$d_{\partial\Omega^{k,t}}(x,y) \le d_{\partial\Omega^{k,0}}(x,y) \le d_\Omega(x)$$

(since $\Omega^{k,0}$ is a cylinder of section Ω) as we wanted to prove. $\square$

Below we return to the usual notation $d = d_{\partial\Omega}$.

Remark 4.1. Observe that $L^p(\Omega, d^{p(1-\beta)})$ contains the set of constant functions if $\beta \le 1$.

Theorem 4.2. *Let Ω be a bounded Hölder α domain, $0 < \alpha \le 1$. Then, for $\alpha \le \beta \le 1$ there exists a constant C depending only on Ω such that*

$$\|f\|_{[L^p(\Omega, d^{p(1-\beta)})]} \le C\|\nabla f\|_{L^p(\Omega, d^{p(1+\alpha-\beta)})^n}. \tag{4.1.2}$$

Proof. Let $B \subset\subset \Omega$ be a ball and φ a regular function supported in B with $\int_B \varphi = 1$. It is enough to prove the result for a function f such that $f_\varphi = 0$.

Let $\omega_0 = \text{dist}(\text{supp }\varphi, \partial\Omega)^t$ and

$$\psi : \Omega^{k,t} \to \mathbf{R} \qquad \psi(x) = \varphi(x)\Pi_{i=1}^k \rho(y_i)$$

where $\rho \in C^\infty[-\omega_0, \omega_0]$ with supp $\rho \subset\subset (-\omega_0, \omega_0)$ and $\int \rho \ne 0$. Then $\psi \in C^\infty(\Omega^{k,t})$ and $\int \psi \ne 0$.

Consider the function $F : \Omega^{k,t} \to \mathbf{R}$ defined by $F(x,y) = f(x)$. Then we have

$$\int_{\Omega^{k,t}} F\psi = \int_B f(x)\varphi(x)dx \left\{\int_{-\omega_0}^{\omega_0} \rho(t)dt\right\}^k = 0.$$

By Lemma C.1 we know that $\Omega^{k,t}$ is Hölder α and then, from Lemma 4.1, it follows that there exists a constant C which depends only on $\Omega^{k,t}$ such that

$$\|F\|_{L^p(\Omega^{k,t})} \leq C\|d^\alpha_{\partial\Omega^{k,t}}\nabla F\|_{L^p(\Omega)}.$$

But,

$$\|F\|^p_{L^p(\Omega^{k,t})} = c_k \int_\Omega f(x)^p d(x)^{tk} = c_k\|d^{\frac{tk}{p}} f\|^p_{L^p(\Omega)}$$

and, using Lemma 4.2, if $1 \leq i \leq n$,

$$\left\|d^\alpha_{\partial\Omega^{k,t}}\frac{\partial F}{\partial x_i}\right\|^p_{L^p(\Omega)} \leq \int_\Omega \frac{\partial f}{\partial x_i} d(x)^{tk+p\alpha} = \|d^{\frac{tk}{p}+\alpha}\frac{\partial f}{\partial x_i}\|^p_{L^p(\Omega)},$$

while if $1 \leq i \leq k$,

$$\left\|d^\alpha_{\partial\Omega^{k,t}}\frac{\partial F}{\partial y_i}\right\|_{L^p(\Omega)} = 0.$$

Therefore, we obtain

$$\|d^{\frac{tk}{p}} f\|_{L^p(\Omega)} \leq C\|d^{\frac{tk}{p}+\alpha}\nabla f\|_{L^p(\Omega)}.$$

The proof concludes by choosing, for example, $k = [p(1-\beta)]+1$ and $t = \frac{ps}{[p(1-\beta)]+1}$.

□

Remark 4.2. The following example shows that an estimate of the form

$$\|d^\mu f\|_{[L^p(\Omega)]} \leq C\|d^\delta\nabla f\|_{L^p(\Omega)^n} \tag{4.1.3}$$

is not valid if $\delta - \mu > \alpha$. Therefore, previous theorem is optimal in this sense.

Given $0 < \alpha < 1$ we call $\gamma = 1/\alpha$ and define the following Hölder α domain:

$$\Omega = \{(x_1,x_2) : 0 < x_1 < 1, -x_1^\gamma < x_2 < x_1^\gamma\},$$

and consider the function

$$f(x_1,x_2) = x_1^{-\nu} - k$$

for some $\nu > 0$ and an arbitrary constant k.

We can easily check that the function $d(x_1,x_2)$ verifies

$$d(x_1,x_2) \sim x_1^\gamma - |x_2|$$

Then, for any $\mu > 0$ and $\delta > 0$ we have that there exists a positive constant C such that

$$\int_\Omega |f|^p d^{\mu p} \geq C^{-1} \int_0^1 x_1^{\frac{\mu p+1}{\alpha} - \nu p} dx_1 \quad \text{and} \quad \int_\Omega |\nabla f|^p d^{\delta p} \leq C \int_0^1 x_1^{\frac{\delta p+1}{\alpha} - (\nu+1)p} dx_1.$$

Then, if $\delta - \mu > \alpha$, we can choose ν such that

$$\frac{\mu}{\alpha} + \frac{1}{p\alpha} + \frac{1}{p} \leq \nu < \frac{\delta}{\alpha} + \frac{1}{p\alpha} + \frac{1}{p} - 1$$

and therefore, for such ν we have

$$\int_\Omega |f|^p d^{\mu p} = \infty \quad \text{and} \quad \int_\Omega |\nabla f|^p d^{\delta p} < \infty.$$

So, it follows that inequality (4.1.3) cannot be true.

Theorem 4.3 (Weighted $Korn^u_{p,B}$ from Weighted Improved Poincaré). *Let Ω be a Hölder α domain and $1 < p < \infty$. Consider a ball $B \subset \Omega$ then, for $\alpha \leq \beta \leq 1$ the following inequality holds,*

$$\|\mathbf{Du}\|_{L^p(\Omega, d^{p(1-\beta)})} \leq C \left\{ \|\boldsymbol{\varepsilon}(\mathbf{u})\|_{L^p(\Omega, d^{p(\alpha-\beta)})} + \|\mathbf{u}\|_{L^p(B)} \right\} \tag{4.1.4}$$

where the constant C depends only on Ω, B, and p.

Proof. Using the same notation and following step by step the third proof of Theorem 3.5, we use Lemma 3.5 to obtain (observe that $\alpha - \beta \leq 0$ and $1 - \beta \geq 0$)

$$\|d^{1-\gamma} \nabla \boldsymbol{\varepsilon}_{ij}(\mathbf{w})\|_{L^p(\Omega)} \leq C \|d^{-\gamma} \boldsymbol{\varepsilon}_{ij}(\mathbf{w})\|_{L^p(\Omega)},$$

instead of (3.2.9). Using (1.0.6),

$$\|d^{1-\gamma} \mathbf{D}(\nabla \mathbf{w}_i)\|_{L^p(\Omega)^{n \times n}} \leq C \|d^{-\gamma} \boldsymbol{\varepsilon}(\mathbf{w})\|_{L^p(\Omega)^{n \times n}}, \tag{4.1.5}$$

and recalling that $\int_\Omega \mathbf{Dw}\phi \, dx = 0$, we get from Theorem 4.2

$$\|d^{1-\beta} \nabla \mathbf{w}_i\|_{L^p(\Omega)^n} \leq C \|d^{\alpha-\beta+1} \mathbf{D} \nabla \mathbf{w}_i\|_{L^p(\Omega)^{n \times n}}$$

and therefore, (4.1.5) with the choice $\gamma = \beta - \alpha$ yields

$$\|d^{1-\beta} \mathbf{Dw}\|_{L^p(\Omega)^{n \times n}} \leq C \|d^{\alpha-\beta} \boldsymbol{\varepsilon}(\mathbf{w})\|_{L^p(\Omega)^{n \times n}}$$

which together with (3.2.7) and (3.2.8) concludes the proof. $\qquad\square$

In the previous chapter we showed that for any Ω, $Korn^u_{p,B} \implies Korn_p$. The same arguments prove that the implication still holds for weighted spaces of the kind considered here.

Corollary 4.2 (Weighted $Korn_p$ from $Korn_{p,B}^u$). *Let $\Omega \subset \mathbb{R}^n$ be a Hölder α domain and $1 < p < \infty$. Then, for $\alpha \leq \beta \leq 1$ the following inequality holds,*

$$\|\mathbf{Du}\|_{[L^p(\Omega,d^{p(1-\beta)})^{n\times n}]_{Skew}} \leq C\|\boldsymbol{\varepsilon}(\mathbf{u})\|_{L^p(\Omega,d^{p(\alpha-\beta)})^{n\times n}} \tag{4.1.6}$$

Proof. Take φ supported in B such that $\int_B \varphi = 1$. Define $\mathbf{x}_\varphi = \int_B \mathbf{x}\varphi\,dx$ and a linear mapping $\mathbf{L}(\mathbf{x}) = \mathbf{v} \in W^{1,p}(\Omega)^n \cap RM$ in the form $\mathbf{v} = \mathbf{u}_\varphi + \mathbf{M}(\mathbf{x} - \mathbf{x}_\varphi)$, with the choice $\mathbf{M} = \boldsymbol{\mu}(\mathbf{u})_B$. Using (4.1.4) with $\mathbf{u} - \mathbf{v}$ we get

$$\|\mathbf{Du}\|_{[L^p(\Omega,d^{p(1-\beta)})^{n\times n}]_{Skew}} \leq C\Big\{\|\boldsymbol{\varepsilon}(\mathbf{u})\|_{L^p(\Omega,d^{p(\alpha-\beta)})^{n\times n}} + \|\mathbf{u} - \mathbf{v}\|_{L^p(B)^n}\Big\}$$

Poincaré and $Korn_p$ inequalities in B, applied to the vector field $\mathbf{u} - \mathbf{v}$, yield

$$\|\mathbf{Du}\|_{[L^p(\Omega,d^{p(1-\beta)})^{n\times n}]_{Skew}} \leq C\Big\{\|\boldsymbol{\varepsilon}(\mathbf{u})\|_{L^p(\Omega,d^{p(\alpha-\beta)})^{n\times n}} + \|\boldsymbol{\varepsilon}(\mathbf{u})\|_{L^p(B)^{n\times n}}\Big\}$$

which implies (4.1.6).   □

4.2 Solutions of the Divergence in Domains with External Cusps

In this section we prove the existence of solutions of the divergence in weighted Sobolev spaces for domains with an external cusp. These results can be used to sharpen the weighted Korn inequality given in the preceding section when the involved domain is an external cusp instead of a general Hölder α domain. For convenience some vectorial variables appear in boldface when they are part of a larger vector.

Given integer numbers $k \geq 1$ and $m \geq 0$ we define

$$\Omega = \Big\{(x,\mathbf{y},\mathbf{z}) \in I \times \mathbb{R}^k \times I^m : |\mathbf{y}| < x^\gamma\Big\} \subset \mathbb{R}^n, \tag{4.2.1}$$

where $n = m+k+1$, I is the interval $(0,1)$ and $\gamma \geq 1$. If $m = 0$, Ω is called a regular cusp, otherwise is called a flat cusp (see Figure 4.1).

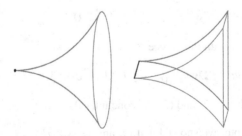

Fig. 4.1 Regular and Flat Cusps

The domain Ω is convex if and only if $\gamma = 1$. The set of singularities of the boundary, which has dimension m, will be called M. Namely,

$$M = \{0\} \times [0,1]^m \subset \mathbb{R}^{k+1} \times \mathbb{R}^m. \tag{4.2.2}$$

We keep the standard notation and the distance to M, defined everywhere in $\mathbb{R}^n$, is called d_M.

Remark 4.3. Notice that for $(x,\mathbf{y},\mathbf{z}) \in \Omega$, $d_M(x,\mathbf{y},\mathbf{z}) \sim x$. Indeed, it is easy to see that $x \leq d_M(x,\mathbf{y},\mathbf{z}) = |(x,\mathbf{y})| \leq (\sqrt{2})x$. In what follows this fact will be repeatedly invoked without further comments.

We need also the following result.

Lemma 4.3. *If* $-(n-m) < \mu < (n-m)(p-1)$, *then* $d_M^\mu \in A_p$.

Proof. Follows straightforwardly from Theorem B.3. $\qquad\square$

In the proof of the main result of this section we will use the Hardy type inequality given in the next lemma.

Lemma 4.4. *Let* Ω *be the domain defined in (4.2.1) and* $1 < p < \infty$. *Given* $\kappa \in \mathbb{R}$, *if* $v \in W_0^{1,p}(\Omega, d_M^{p\kappa})$, *then* $v/x \in L^p(\Omega, d_M^{p\kappa})$ *and there exists constant* C, *depending only on* p, *such that*

$$\left\| \frac{v}{x} \right\|_{L^p(\Omega, d_M^{p\kappa})} \leq C \left\| \frac{\partial v}{\partial y_1} \right\|_{L^p(\Omega, d_M^{p\kappa})}. \tag{4.2.3}$$

Consequently, $W_0^{1,p}(\Omega, d_M^{p\kappa})$ *is continuously embedded in* $W_0^{1,p}(\Omega, d_M^{p(\kappa-1)}, d_M^{p\kappa})$.

Proof. By density it is enough to prove (4.2.3) for $v \in C_0^\infty(\Omega)$. Writing $x^{p\kappa-p} = \frac{\partial(x^{p\kappa-p}y_1)}{\partial y_1}$ and integrating by parts we have

$$\int_\Omega |v(x,\mathbf{y},\mathbf{z})|^p x^{p\kappa-p} dx\,d\mathbf{y}\,d\mathbf{z} = \int_\Omega |v(x,\mathbf{y},\mathbf{z})|^p \frac{\partial(x^{p\kappa-p}y_1)}{\partial y_1} dx\,d\mathbf{y}\,d\mathbf{z}$$

$$= -\int_\Omega \frac{\partial(|v(x,\mathbf{y},\mathbf{z})|^p)}{\partial y_1} x^{p\kappa-p}y_1 \, dx\,d\mathbf{y}\,d\mathbf{z}$$

Then, since $|y_1| \leq x$, we have

$$\int_\Omega |v(x,\mathbf{y},\mathbf{z})|^p x^{p\kappa-p} dx\,d\mathbf{y}\,d\mathbf{z} \leq p\int_\Omega |v(x,\mathbf{y},\mathbf{z})|^{p-1} \left| \frac{\partial v(x,\mathbf{y},\mathbf{z})}{\partial y_1} \right| x^{p\kappa-p+1} dx\,d\mathbf{y}\,d\mathbf{z}$$

and so, writing now $x^{p\kappa-p+1} = x^{\frac{p\kappa-p}{p'}} x^\kappa$ and applying the Hölder inequality, we obtain

$$\int_\Omega |v(x,\mathbf{y},\mathbf{z})|^p x^{p\kappa-p} dx\,d\mathbf{y}\,d\mathbf{z}$$

$$\leq p \left(\int_\Omega |v(x,\mathbf{y},\mathbf{z})|^p x^{p\kappa-p} \, dx\,d\mathbf{y}\,d\mathbf{z} \right)^{(p-1)/p} \left(\int_\Omega \left| \frac{\partial v(x,\mathbf{y},\mathbf{z})}{\partial y_1} \right|^p x^{p\kappa} \, dx\,d\mathbf{y}\,d\mathbf{z} \right)^{1/p}$$

and therefore,

$$\left(\int_{\Omega}|v(x,\mathbf{y},\mathbf{z})|^p x^{p\kappa-p}dx\,d\mathbf{y}\,d\mathbf{z}\right)^{1/p}\leq p\left(\int_{\Omega}\left|\frac{\partial v(x,\mathbf{y},\mathbf{z})}{\partial y_1}\right|^p x^{p\kappa}\,dx\,d\mathbf{y}\,d\mathbf{z}\right)^{1/p}.$$

To conclude the proof we use that $d_M(x,\mathbf{y},\mathbf{z})\simeq x$ and then, that $W_0^{1,p}(\Omega,d_M^{p\kappa})$ is continuously embedded in $W_0^{1,p}(\Omega,d_M^{p(\kappa-1)},d_M^{p\kappa})$ follows from (4.2.3). $\square$

We can now prove the main result of this section.

Theorem 4.4. *Let Ω be the domain defined in (4.2.1) for a fixed $\gamma > 1$, M defined as in (4.2.2), and $1 < p < \infty$. If*

$$\beta \in \left(\frac{-\gamma(n-m)}{p}-\frac{\gamma-1}{p'},\frac{\gamma(n-m)}{p'}-\frac{\gamma-1}{p'}\right)$$

and $\eta \in \mathbb{R}$ is such that $\eta \geq \beta + \gamma - 1$, then given $f \in L_0^p(\Omega,d_M^{p\beta})$, there exists $\mathbf{u} \in W_0^{1,p}(\Omega,d_M^{p(\eta-1)},d_M^{p\eta})^n$ satisfying

$$\operatorname{div}\mathbf{u} = f \tag{4.2.4}$$

and

$$\|\mathbf{u}\|_{W^{1,p}(\Omega,d_M^{p(\eta-1)},d_M^{p\eta})^n} \leq C\|f\|_{L^p(\Omega,d_M^{p\beta})} \tag{4.2.5}$$

with a constant C depending only on γ, β, η, p, and n.

Proof. It is enough to prove the result for the case $\eta = \beta + \gamma - 1$ and therefore we consider only this case.

Define the *convex* domain

$$\widehat{\Omega} = \left\{(\hat{x},\hat{\mathbf{y}},\hat{\mathbf{z}}) \in I \times \mathbb{R}^k \times I^m : |\hat{\mathbf{y}}| < \hat{x}\right\} \subset \mathbb{R}^n \tag{4.2.6}$$

and let $\mathbf{F} : \widehat{\Omega} \to \Omega$ be the one-to-one application given by

$$\mathbf{F}(\hat{x},\hat{\mathbf{y}},\hat{\mathbf{z}}) = (\hat{x}^\alpha,\hat{\mathbf{y}},\hat{\mathbf{z}}) = (x,\mathbf{y},\mathbf{z}),$$

where $\alpha = 1/\gamma$.

By this change of variables we associate functions defined in Ω with functions defined in $\widehat{\Omega}$ in the following way,

$$h(x,\mathbf{y},\mathbf{z}) = \hat{h}(\hat{x},\hat{\mathbf{y}},\hat{\mathbf{z}}).$$

Now, for $f \in L_0^p(\Omega,d_M^{p\beta})$, we define $\hat{g} : \widehat{\Omega} \to \Omega$ by

$$\hat{g}(\hat{x},\hat{\mathbf{y}},\hat{\mathbf{z}}) := \alpha\hat{x}^{\alpha-1}\hat{f}(\hat{x},\hat{\mathbf{y}},\hat{\mathbf{z}}).$$

We want to apply Theorem 2.8 for $\hat{g}$ on the convex domain $\widehat{\Omega}$ and then obtain the desired solution of (4.2.4) by using the so-called Piola transform for vector fields [24].

In the rest of the proof we use several times the equivalence mentioned in Remark 4.3, together with the following facts $\det \mathbf{DF}(\hat{x}, \hat{y}, \hat{z}) = \alpha \hat{x}^{\alpha-1}$ and $\det \mathbf{DF}^{-1}(x, y, z) = \gamma x^{\gamma-1}$.

First let us see that, for $\hat{\beta} = \alpha (\beta + (\gamma - 1)/p')$, we have

$$\hat{g} \in L_0^p(\widehat{\Omega}, d_M^{p\hat{\beta}}) \qquad \text{and} \qquad \|\hat{g}\|_{L^p(\widehat{\Omega}, d_M^{p\hat{\beta}})} \simeq \|f\|_{L^p(\Omega, d_M^{p\beta})}. \qquad (4.2.7)$$

Indeed, we have

$$\|\hat{g}\|_{L^p(\widehat{\Omega}, d_M^{p\hat{\beta}})}^p \simeq \int_{\widehat{\Omega}} |\hat{g}|^p \hat{x}^{p\hat{\beta}} = \alpha^p \int_{\widehat{\Omega}} |\hat{f}|^p \hat{x}^{p(\alpha-1)} \hat{x}^{\alpha p(\beta+(\gamma-1)/p')}$$

$$= \alpha^p \int_{\Omega} |f|^p x^{p\beta+1-\gamma} \gamma x^{\gamma-1} \simeq \|f\|_{L^p(\Omega, d_M^{p\beta})}^p$$

and

$$\int_{\widehat{\Omega}} \hat{g} = \alpha \int_{\widehat{\Omega}} \hat{f} \hat{x}^{\alpha-1} = \alpha \int_{\Omega} f x^{1-\gamma} \gamma x^{\gamma-1} = \int_{\Omega} f = 0.$$

Thus, (4.2.7) holds.

Observe that, from Lemma (4.3) and our hypothesis on β, we have $d_M^{p\hat{\beta}} \in A_p$. In particular, it follows that $\hat{g} \in L^1(\widehat{\Omega})$ and therefore the mean value of f in Ω is well defined.

Now, from Theorem 2.8 we know that there exists $\hat{\mathbf{v}} \in W_0^{1,p}(\widehat{\Omega}, d_M^{p\hat{\beta}})^n$ such that

$$\operatorname{div} \hat{\mathbf{v}} = \hat{g} \qquad (4.2.8)$$

and

$$\|\hat{\mathbf{v}}\|_{W^{1,p}(\widehat{\Omega}, d_M^{p\hat{\beta}})^n} \leq C \|\hat{g}\|_{L^p(\widehat{\Omega}, d_M^{p\hat{\beta}})}. \qquad (4.2.9)$$

Now, we define $\mathbf{u}$ as the Piola transform of $\hat{\mathbf{v}}$, namely,

$$\mathbf{u}(x, y, z) = \frac{1}{\det \mathbf{DF}} \mathbf{DF}(\hat{x}, \hat{y}, \hat{z}) \hat{\mathbf{v}}(\hat{x}, \hat{y}, \hat{z})$$

or equivalently, if $\hat{\mathbf{v}} = (\hat{v}_1, \ldots, \hat{v}_n)$,

$$\mathbf{u}(x, y, z) = \gamma x^{\gamma-1} \left(\alpha x^{1-\gamma} \hat{v}_1(x^\gamma, y, z), \hat{v}_2(x^\gamma, y, z), \ldots, \hat{v}_n(x^\gamma, y, z) \right).$$

Then, using (4.2.8), it is easy to see that

$$\operatorname{div} \mathbf{u} = f.$$

To prove (4.2.5) we first show that

$$\|\mathbf{u}\|_{W^{1,p}(\Omega, d_M^{p\eta})^n} \leq C \|f\|_{L^p(\Omega, d_M^{p\beta})}. \qquad (4.2.10)$$

In view of the equivalence of norms given in (4.2.7) and the estimate (4.2.9), to prove (4.2.10) it is enough to see that

$$\|\mathbf{u}\|_{W^{1,p}(\Omega,d_M^{p\eta})^n} \leq C\|\hat{\mathbf{v}}\|_{W^{1,p}(\hat{\Omega},d_M^{p\hat\beta})^n}. \tag{4.2.11}$$

But, we have

$$\|u_1\|^p_{L^p(\Omega,d_M^{p\eta})} \simeq \int_\Omega |u_1|^p x^{p\eta} = \alpha \int_{\hat\Omega} |\hat{v}_1|^p \hat{x}^{\alpha p\eta}\hat{x}^{\alpha-1} \simeq \|\hat{v}_1\|^p_{L^p(\Omega,d_M^{p\hat\beta})}, \tag{4.2.12}$$

where in the last step we have used $\alpha p\eta + \alpha - 1 = p\hat\beta$. In an analogous way we can show that, for $j=2,\dots,n$,

$$\|u_j\|_{L^p(\Omega,d_M^{p\eta})} \leq C\|\hat{v}_j\|_{L^p(\Omega,d_M^{p\hat\beta})}.$$

Then, it only remains to bound the derivatives of the components of $\mathbf{u}$. That

$$\left\|\frac{\partial u_1}{\partial y_1}\right\|^p_{L^p(\Omega,d_M^{p\eta})} \simeq \left\|\frac{\partial\hat{v}_1}{\partial\hat{y}_1}\right\|^p_{L^p(\hat\Omega,d_M^{p\hat\beta})}$$

follows exactly as (4.2.12). Let us now estimate $\frac{\partial u_2}{\partial x}$. Using

$$\left|\frac{\partial u_2}{\partial x}\right| = \gamma^2\left|\frac{\gamma-1}{\gamma}\frac{\hat{v}_2(x^\gamma,\mathbf{y},\mathbf{z})}{x^\gamma} + \frac{\partial\hat{v}_2(x^\gamma,\mathbf{y},\mathbf{z})}{\partial\hat{x}}\right| x^{2(\gamma-1)}$$

and Lemma 4.4 for $\hat\Omega$ we have

$$\left\|\frac{\partial u_2}{\partial x}\right\|^p_{L^p(\Omega,d_M^{p\eta})} \simeq \int_\Omega \left|\frac{\partial u_2}{\partial x}\right|^p x^{p\eta}$$
$$\leq C\int_{\hat\Omega}\left(\left|\frac{\hat{v}_2}{\hat{x}}\right|^p + \left|\frac{\partial\hat{v}_2}{\partial\hat{x}}\right|^p\right)\hat{x}^{2p(1-\alpha)}\hat{x}^{\alpha p\eta+\alpha-1}$$
$$\leq C\int_{\hat\Omega}\left(\left|\frac{\hat{v}_2}{\hat{x}}\right|^p + \left|\frac{\partial\hat{v}_2}{\partial\hat{x}}\right|^p\right)\hat{x}^{p\hat\beta}$$
$$\leq C\int_{\hat\Omega}\left|\frac{\partial\hat{v}_2}{\partial\hat{x}}\right|^p\hat{x}^{p\hat\beta} = \left\|\frac{\partial\hat{v}_2}{\partial\hat{x}}\right\|^p_{L^p(\hat\Omega,d_M^{p\hat\beta})},$$

where we have used again $\alpha p\eta + \alpha - 1 = p\hat\beta$ and that $2p(1-\alpha) > 0$.

All the other derivatives of the components of $\mathbf{u}$ can be bounded in an analogous way and therefore (4.2.11) holds.

Now, since

$$\mathbf{u}|_{\partial\Omega} = \frac{1}{\det\mathbf{DF}}\mathbf{DF}\hat{\mathbf{v}}|_{\partial\hat\Omega},$$

it is easy to check that $\mathbf{u}$ belongs to the closure of $C_0^\infty(\Omega)^n$, i.e., $\mathbf{u} \in W_0^{1,p}(\Omega,d_M^{p\eta})^n$ and by Lemma 4.4 $\mathbf{u} \in W_0^{1,p}(\Omega,d_M^{p(\eta-1)},d_M^{p\eta})^n$ as we wanted to show. $\square$

Remark 4.4. Notice that the condition $\beta < \frac{\gamma(n-m)}{p'} - \frac{\gamma-1}{p'}$ implies $L^p(\Omega, d_M^{p\beta}) \subset L^1(\Omega)$, and therefore, $L_0^p(\Omega, d_M^{p\beta})$ is well defined, while if $\beta \geq \frac{\gamma(n-m)}{p'} - \frac{\gamma-1}{p'}$, it is easy to check that $f(x, \mathbf{y}, \mathbf{z}) = (1 - \log x)^{-1} x^{\gamma-1-\gamma(n-m)}$ belongs to $L^p(\Omega, d_M^{p\beta}) \setminus L^1(\Omega)$.

Remark 4.5. The range for η in Theorem 4.4 is sharp in the sense that if $\eta - \beta < \gamma - 1$, there exists $f \in L_0^p(\Omega, d_M^{p\beta})$ such that a solution $\mathbf{u}$ of (4.2.4) satisfying (4.2.5) does not exist. See Section 4.4.

4.3 Weighted Korn Inequalities on External Cusps

Since in a Hölder α domain singularities may arise arbitrarily along the boundary while in a standard external cusp there is a single and well-located singular point it is expected an improved version of Theorem 4.3 using d_M instead of d. Actually, in the light of the previous section we can easily derive this result. First, we use Lemma 3.6 to get the following.

Theorem 4.5 (Weighted Korn from div$_p$ in Weighted Spaces). *Let ω be a weight in a bounded domain $\Omega \subset \mathbb{R}^n$. Assume that for any $1 < p < \infty$, and any $g \in L_0^{p'}(\Omega)$ there exists $\mathbf{u} \in W_0^{1,p'}(\Omega, \omega^{1-p'})^n$ such that $\operatorname{div} \mathbf{u} = g$ with*

$$\|\mathbf{u}\|_{W^{1,p'}(\Omega, \omega^{1-p'})^n} \leq C\|g\|_{L^{p'}(\Omega)},$$

with a constant C depending only on Ω, p, and ω. Fix an open ball $B \subset \Omega$. Then, for any $\mathbf{v} \in W^{1,p}(\Omega)^n$,

$$\|\mathbf{Dv}\|_{L^p(\Omega)^{n\times n}} \leq C\left\{\|\mathbf{v}\|_{L^p(B)^n} + \|\boldsymbol{\varepsilon}(\mathbf{v})\|_{L^p(\Omega, \omega)^{n\times n}}\right\},$$

where the constant C depends only on Ω, B, p, and ω.

Proof. Applying Lemma 3.6 to each $\frac{\partial v_i}{\partial x_j}$

$$\left\|\frac{\partial v_i}{\partial x_j}\right\|_{L^p(\Omega)} \leq C\left\{\left\|\frac{\partial v_i}{\partial x_j}\right\|_{W^{-1,p}(B)} + \left\|\nabla \frac{\partial v_i}{\partial x_j}\right\|_{W^{-1,p}(\Omega, \omega)^n}\right\}.$$

which together with (1.0.6) in the last term on the right-hand side yields the desired result.   □

An immediate consequence of Theorems 4.5 and 4.4 is the following.

Corollary 4.3 (Weighted Korn for External Cusps). *Given $\gamma \geq 1$, let Ω be the domain defined in (4.2.1), M defined in (4.2.2), $1 < p < \infty$, and $B \subset \Omega$ an open ball. Then, there exists a constant C, which depends only on Ω, B, and p, such that for all $\mathbf{u} \in W^{1,p}(\Omega)$,*

$$\|D\mathbf{u}\|_{L^p(\Omega)^{n\times n}} \leq C\left\{\|\boldsymbol{\varepsilon}(\mathbf{u})\|_{L^p(\Omega,d_M^{p(1-\gamma)})^{n\times n}} + \|\mathbf{u}\|_{L^p(B)^n}\right\}.$$

Proof. According to Theorem 4.4, for any $g \in L_0^{p'}(\Omega)$ there exists $\mathbf{u} \in W_0^{1,p'}$ $(\Omega, d_M^{p'(\gamma-1)})^n$ such that div $\mathbf{u} = g$ and

$$\|\mathbf{u}\|_{W^{1,p'}(\Omega,d_M^{p'(\gamma-1)})^n} \leq C\|g\|_{L^{p'}(\Omega)},$$

with a constant C depending only on γ and p. Therefore, Theorem 4.5 applies for $\omega = d_M^{p(1-\gamma)}$.   □

More general Korn type inequalities for those domains defined in (4.2.1) can be obtained using an argument introduced in [17]. Essentially a weight can be added to each side of Korn's inequality involving derivatives.

Theorem 4.6 (Weighted Korn$_{p,B}^u$ for External Cusps). *Given $\gamma \geq 1$, let Ω be the domain defined in (4.2.1), M defined in (4.2.2), $1 < p < \infty$, $B \subset \Omega$ an open ball, and $\beta \geq 0$. Then, there exists a constant C, which depends only on Ω, B, p, and β, such that for all $\mathbf{u} \in W^{1,p}(\Omega, d_M^{p\beta})$*

$$\|D\mathbf{u}\|_{L^p(\Omega,d_M^{p\beta})^{n\times n}} \leq C\left\{\|\boldsymbol{\varepsilon}(\mathbf{u})\|_{L^p(\Omega,d_M^{p(\beta+1-\gamma)})^{n\times n}} + \|\mathbf{u}\|_{L^p(B)^n}\right\}.$$

Proof. To simplify notation we will assume that $m = 0$ in the definition of Ω. The other cases can be treated analogously.

Let $n' \in \mathbb{N}_0$ and $0 < s \leq \gamma$ be such that $sn' = p\beta$. As in [17] we introduce

$$\Omega^{n',s} = \{(x,\mathbf{y},\mathbf{z}') \in \mathbb{R}^{n+n'} : (x,\mathbf{y}) \in \Omega, \mathbf{z}' \in \mathbb{R}^{n'} \text{ with } |\mathbf{z}'| < x^s\}. \quad (4.3.1)$$

Suppose that the hypothesis in Theorem 4.5 on solutions of the divergence is verified for $\Omega = \Omega^{n',s}$ and $\omega = x^{p(1-\gamma)}$. Then, if $B' \subset \Omega^{n',s}$ is a ball with the same radius and center than B, we have, calling $m = n + n'$,

$$\|D\mathbf{v}\|_{L^p(\Omega^{n',s})^{m\times m}} \leq C\Big\{\|\mathbf{v}\|_{L^p(B')^m}$$
$$+ \|\boldsymbol{\varepsilon}(\mathbf{v})\|_{L^p(\Omega^{n',s},x^{p(1-\gamma)})^{m\times m}}\Big\}, \quad (4.3.2)$$

for all $\mathbf{v} \in W^{1,p}(\Omega^{n',s})^m$.

Now, given $\mathbf{u}$ in $W^{1,p}(\Omega, d_M^{p\beta})^n$ we define

$$\mathbf{v}(x,y,\mathbf{z}') = (\mathbf{u}(x,y),\underbrace{0,\ldots,0}_{n'}).$$

Then, using that for $(x,\mathbf{y}) \in \Omega$, $d_M(x,\mathbf{y}) \simeq x$, it is easy to check that (4.3.2) is equivalent to

$$\|D\mathbf{u}\|_{L^p(\Omega,d_M^{p\beta})^{n\times n}} \leq C\left\{\|\mathbf{u}\|_{L^p(B)^n} + \|\boldsymbol{\varepsilon}(\mathbf{u})\|_{L^p(\Omega,d_M^{p(\beta+1-\gamma)})^{n\times n}}\right\}.$$

Hence, to finish the proof we have to verify the hypothesis of Theorem 4.5 for the domain $\Omega^{n',s}$ with the weight $\omega = x^{p(1-\gamma)}$. Since in this case $\omega^{1-p'} = x^{p'(\gamma-1)}$, we have to show that, for any $g \in L_0^{p'}(\Omega^{n',s})$, there exists $\mathbf{w} \in W_0^{1,p'}(\Omega^{n',s}, x^{p'(\gamma-1)})^n$ such that div $\mathbf{w} = g$ and

$$\|\mathbf{w}\|_{W^{1,p'}(\Omega^{n',s}, x^{p'(\gamma-1)})^n} \le C\|g\|_{L^{p'}(\Omega^{n',s})}.$$

But this can be proved exactly as Theorem 4.4, using now the convex domain

$$\widehat{\Omega}^{n',s} := \{(\hat{x}, \hat{\mathbf{y}}, \hat{\mathbf{z}}') \in \mathbb{R}^{n+n'} : (\hat{x}, \hat{\mathbf{y}}) \in \widehat{\Omega}, \mathbf{z}' \in \mathbb{R}^{n'} \text{ with } |\mathbf{z}'| < x^{\alpha s}\},$$

with $\widehat{\Omega}$ defined as in (4.2.6), and the one-to-one map $F : \widehat{\Omega}^{n',s} \to \Omega^{n',s}$ defined by

$$F(\hat{x}, \hat{\mathbf{y}}, \hat{\mathbf{z}}') := (\hat{x}^{\alpha}, \hat{\mathbf{y}}, \hat{\mathbf{z}}'). \qquad \Box$$

Remark 4.6. For $\gamma > 1$, call $\alpha = 1/\gamma$ and consider the following Hölder α domain (see Figure 4.2)

$$\Omega := \{(x,y) \in \mathbb{R}^2 : 0 < x < 1, |y| < x^{\gamma}\}. \qquad (4.3.3)$$

Notice that, for such an Ω, Theorem 4.6 with $\beta = \gamma - 1$ is stronger than Theorem 4.3 with $\beta = \alpha$. Indeed, for $(x,y) \in \Omega$, we have $d_M(x,y) = \|(x,y)\| \simeq x$ and then $d^{1-\alpha}(x,y) \le x^{\gamma(1-\alpha)} \simeq d_M^{\gamma-1}(x,y)$ while $d_M^{\gamma-1}(x,y)$ cannot be bounded by positive powers of $d(x,y)$.

It is possible to deal with more general cusps obtaining similar results as those given in this section. This is treated later in Section 4.5.

4.4 Some Simple Counterexamples

Following [2] we present very simple counterexamples for both problems Korn_p and div_p for cuspidal domains. As it is shown below, these counterexamples can also be used to prove optimality of the results in weighted norms obtained so far. For the sake of clarity we present first the results in two dimensions and explain afterwards how they can be extended for higher dimensional domains.

Let Ω be defined by (4.3.3) and the vector field

$$\mathbf{w} = (u,v) = ((s-1)yx^{-s}, x^{1-s}),$$

with $s \in \mathbb{R}$, $s \neq 1$, to be chosen below.

Recall that Ω is not a Lipschitz domain but it is Hölder α, with $\alpha = \frac{1}{\gamma}$.
We have

$$\mathbf{Dw} = \begin{pmatrix} -s(s-1)yx^{-s-1} & (s-1)x^{-s} \\ (1-s)x^{-s} & 0 \end{pmatrix}$$

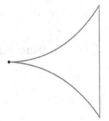

Fig. 4.2 2-Dimensional external cusp

while, on the other hand,

$$\boldsymbol{\varepsilon}(\mathbf{w}) = \begin{pmatrix} -s(s-1)yx^{-s-1} & 0 \\ 0 & 0 \end{pmatrix}.$$

A straightforward computation shows that

$$\|\mathbf{w}\|_{L^p(\Omega)}^p \le C \left(\int_0^1 \int_0^{x^\gamma} (y^p x^{-ps} + x^{p(1-s)})dydx \right)$$

$$\le C \left(\int_0^1 \int_0^{x^\gamma} x^{p(1-s)} dydx \right) = C \int_0^1 x^{p-ps+\gamma} dx,$$

and

$$\|\boldsymbol{\varepsilon}(\mathbf{w})\|_{L^p(\Omega)}^p \le C \int_0^1 \int_0^{x^\gamma} y^p x^{-p(s+1)} dydx \le C \int_0^1 x^{\gamma(p+1)-p(1+s)} dx,$$

hence

$$s < \min \left\{ \frac{(\gamma+1)}{p} + (\gamma-1), \frac{(\gamma+1)}{p} + 1 \right\} \Rightarrow \|\boldsymbol{\varepsilon}(\mathbf{w})\|_{L^p(\Omega)^p}, \|\mathbf{w}\|_{L^p(\Omega)^p} < \infty.$$
$$(4.4.1)$$

However, we have

$$\left\| \frac{\partial u}{\partial y} \right\|_{L^p(\Omega)}^p = C \int_0^1 x^{-sp+\gamma} dx,$$

and so,

$$\left\| \frac{\partial u}{\partial y} \right\|_{L^p(\Omega)} < \infty \Rightarrow s < \frac{\gamma+1}{p}. \qquad (4.4.2)$$

But, since $\gamma > 1$, it is possible to take s such that

$$\frac{\gamma+1}{p} \le s < \min \left\{ \frac{(\gamma+1)}{p} + (\gamma-1), \frac{(\gamma+1)}{p} + 1 \right\},$$

and therefore, it follows from (4.4.1) and (4.4.2) that $Korn_p^u$ inequality cannot be valid in Ω. Moreover, let us show that the same counterexample leads to the opti-mality of several results in weighted spaces.

For example, consider Theorem 4.3 with $\alpha = \beta$. Then for $\Omega \subset \mathbb{R}^n$ a Hölder α domain in any space dimension $n \geq 2$, the following weighted inequality holds for any $1 < p < \infty$,

$$\|d^{1-\alpha}\mathbf{D}\mathbf{w}\|_{L^p(\Omega)} \leq C\{\|\boldsymbol{\varepsilon}(\mathbf{w})\|_{L^p(\Omega)} + \|\mathbf{w}\|_{L^p(B)}\}, \tag{4.4.3}$$

where B is a fixed ball such that $B \subset \Omega$. The same field $\mathbf{w}$ used in our counterexample for the classical inequality shows that (4.4.3) is sharp in the sense that the power of d appearing on the left-hand side cannot be reduced. Indeed, since the last term on the right-hand side is the norm of $\mathbf{w}$ on the ball B, the same computations given above give now,

$$s < \frac{(\gamma+1)}{p} + (\gamma-1) \Rightarrow \|\boldsymbol{\varepsilon}(\mathbf{w})\|_{L^p(\Omega)}, \|\mathbf{w}\|_{L^p(B)} < \infty \tag{4.4.4}$$

instead of (4.4.1).

Introducing the subdomain $\Omega_{\frac{1}{2}} \subset \Omega$

$$\Omega_{\frac{1}{2}} := \left\{(x,y) \in \mathbb{R}^2 : 0 < x < 1, \ 0 < |y| < \frac{1}{2}x^\gamma\right\}, \tag{4.4.5}$$

we have, for any β,

$$\left\|d^\beta \frac{\partial u}{\partial y}\right\|_{L^p(\Omega)}^p \geq \left\|d^\beta \frac{\partial u}{\partial y}\right\|_{L^p\left(\Omega_{\frac{1}{2}}\right)}^p.$$

Now, it can be easily checked that for this Ω, $d(x,y) \sim x^\gamma - |y|$ and therefore, for $(x,y) \in \Omega_{\frac{1}{2}}$ we obviously have $d(x,y) \sim x^\gamma - |y| \geq \frac{1}{2}x^\gamma$.

Hence

$$\left\|d^\beta \frac{\partial u}{\partial y}\right\|_{L^p(\Omega)}^p \geq C \int_0^1 \int_0^{x^\gamma/2} x^{\beta\gamma p - sp} dy dx = C \int_0^1 x^{\beta\gamma p - sp + \gamma} dx$$

and so, $\|d^\beta \frac{\partial u}{\partial y}\|_{L^p(\Omega)}$ is not finite if $\frac{\gamma+1}{p} + \beta\gamma \leq s$. On the other hand, for any $\beta < 1 - \alpha = 1 - \frac{1}{\gamma}$ it is possible to take s such that

$$\frac{\gamma+1}{p} + \beta\gamma \leq s < \frac{(\gamma+1)}{p} + (\gamma-1)$$

which, in view of (4.4.4), shows the optimality of (4.4.3) in the sense that it is not possible to replace the power $1 - \alpha$ on the left-hand side by any power $\beta < 1 - \alpha$.

On the other hand, for Ω as in (4.3.3) Theorem 4.6 yields with $\beta = \gamma - 1$

$$\|d_M^{\gamma-1}\mathbf{D}\mathbf{w}\|_{L^p(\Omega)} \leq C\{\|\boldsymbol{\varepsilon}(\mathbf{w})\|_{L^p(\Omega)} + \|\mathbf{w}\|_{L^p(B)}\}, \tag{4.4.6}$$

where B is a fixed ball and d_M the distance to the cusp placed at $(0,0)$.

It follows immediately that (4.4.6) cannot be improved by taking on the left-hand side a smaller power of d_M (see Remark 4.6). Indeed, since for any $\beta > 0$, $d^{\alpha\beta} \le d_M^\beta$ in Ω, if we could replace $\gamma - 1$ by some $\beta < \gamma - 1$ in (4.4.6), we could also improve (4.4.3) by replacing $\alpha - 1$ by $\alpha\beta < 1 - \alpha$.

We have considered the particular cases (4.4.3) and (4.4.6) in order to simplify technical details. However, a simple generalization of the vector field introduced for our counterexamples can be used to obtain similar results in arbitrary higher dimensions. Indeed, let us define now,

$$\Omega := \left\{ (x, \mathbf{y}) \in (0,1) \times \mathbb{R}^{n-1} : \|\mathbf{y}\| < x^\gamma \right\}. \tag{4.4.7}$$

Taking $\mathbf{w} = ((s-1)(y_1 + y_2 + \cdots y_{n-1})x^{-s}, x^{1-s}, \cdots, x^{1-s})$ we have

$$D_{11}\mathbf{w} = -s(s-1)(y_1 + \cdots + y_{n-1})x^{-s-1},$$

$$D_{1j}\mathbf{w} = -D_{j1}\mathbf{w} = (s-1)x^{-s} \quad \text{for} \quad j > 1$$

and $D_{ij}\mathbf{w} = 0$ otherwise. Therefore, the only nonzero component of $\boldsymbol{\varepsilon}(\mathbf{w})$ is

$$\varepsilon_{11}(\mathbf{w}) = -s(s-1)(y_1 + \cdots + y_{n-1})x^{-s-1}.$$

Then, a straightforward calculation yields

$$\|\boldsymbol{\varepsilon}(\mathbf{w})\|_{L^p(\Omega)}^p = C \int_\Omega |y_1 + \cdots + y_{n-1}|^p x^{-p(s+1)} dy dx$$

$$\le C \int_\Omega \|\mathbf{y}\|^p x^{-p(s+1)} dy dx$$

$$\le C \int_0^1 \int_0^{x^\gamma} \rho^{p+n-2} x^{-p(s+1)} d\rho dx$$

$$\le C \int_0^1 x^{\gamma(p+n-1)-p(s+1)} dx,$$

$$\|\mathbf{w}\|_{L^p(\Omega)}^p \le C \int_\Omega x^{(1-s)p} dy dx \le C \int_0^1 x^{\gamma(n-1)+(1-s)p} dx$$

and

$$\|D\mathbf{w}\|_{L^p(\Omega)}^p \ge C \int_\Omega x^{-sp} dy dx \ge C \int_0^1 x^{\gamma(n-1)-sp} dx.$$

Then, if

$$\frac{\gamma(n-1)+1}{p} \le s < \min\left\{ \frac{\gamma(n-1)+1}{p} + \gamma - 1, \frac{\gamma(n-1)+1}{p} + 1 \right\}, \tag{4.4.8}$$

$\|\boldsymbol{\varepsilon}(\mathbf{w})\|_{L^p(\Omega)}$ and $\|\mathbf{w}\|_{L^p(\Omega)}$ are finite while $\|D\mathbf{w}\|_{L^p(\Omega)}$ is not. But, since $\gamma > 1$, it is possible to choose s satisfying (4.4.8), and therefore, (3.1.3) does not hold in Ω.

Moreover, similar arguments can be applied to show that $Korn_p^u$ is not valid in more general cusps. Indeed, for $k \geq 1$, define

$$\Omega := \left\{ (x, \mathbf{y}, \mathbf{z}) \in (0,1) \times \mathbb{R}^k \times (0,1)^{n-k-1} \; : \; \|\mathbf{y}\| < x^\gamma \right\}. \qquad (4.4.9)$$

We can generalize our counterexample by taking

$$\mathbf{w} = ((s-1)(y_1 + y_2 + \cdots y_k)x^{-s}, \overbrace{x^{1-s}, \cdots, x^{1-s}}^{k}, \overbrace{0, \cdots, 0}^{n-k-1}).$$

Now we have

$$\|\boldsymbol{\varepsilon}(\mathbf{w})\|_{L^p(\Omega)}^p = C \int_\Omega |y_1 + \cdots + y_k|^p x^{-p(s+1)} d\mathbf{y} dx d\mathbf{z}$$

$$\leq C \int_\Omega \|\mathbf{y}\|^p x^{-p(s+1)} d\mathbf{y} dx d\mathbf{z}$$

$$\leq C \int_0^1 \int_0^{x^\gamma} \rho^{p+k-1} x^{-p(s+1)} d\rho dx$$

$$\leq C \int_0^1 x^{\gamma(p+k)-p(s+1)} dx,$$

$$\|\mathbf{w}\|_{L^p(\Omega)}^p \leq C \int_\Omega x^{(1-s)p} d\mathbf{y} dx \leq C \int_0^1 x^{\gamma k + (1-s)p} dx,$$

and

$$\|\mathbf{Dw}\|_{L^p(\Omega)}^p \geq \int_\Omega x^{-sp} d\mathbf{y} dx \geq C \int_0^1 x^{\gamma k - sp} dx.$$

Therefore, taking now s such that

$$\frac{\gamma k + 1}{p} \leq s < \min \left\{ \frac{\gamma k + 1}{p} + \gamma - 1, \frac{\gamma k + 1}{p} + 1 \right\},$$

which is possible for any $\gamma > 1$, we conclude that (3.1.3) is not valid for the class of domains defined in (4.4.9).

Let us mention that the vector fields introduced for the counterexamples can be used to show the optimality of the powers in the weighted estimates (4.4.3) and (4.4.6) for the domains (4.4.7) and (4.4.9) (in this last case, and for (4.4.6), d_M is defined as the distance to the singular set of the boundary placed at $(0,0,\mathbf{z})$, $0 < \mathbf{z} < 1$).

Since for unweighted spaces div_p implies $Korn_p$, the given counterexamples immediately show that the involved cuspidal domains do not satisfy div_p. This remark also applies in some particular weighted cases, nevertheless, it is not difficult to see by direct computations that the same vector fields introduced above can be used to show that cuspidal domains, such as those given in (4.4.7) and (4.4.9), do not satisfy div_p. Moreover, as for the weighted Korn inequalities, it is possible to show the optimality of the weighted versions of div_p obtained in Theorem 4.4.

Let us show that the condition $\eta \geq \beta + \gamma - 1$ in Theorem 4.4 is sharp. As in that theorem, we assume that

$$\beta \in \left(\frac{-\gamma(n-m)}{p} - \frac{\gamma-1}{p'}, \frac{\gamma(n-m)}{p'} - \frac{\gamma-1}{p'} \right).$$

We will also assume that constant functions belong to $L^p(\Omega, d_M^{p\beta})$ which leads to the condition

$$0 < \frac{\gamma k + 1}{p} + \beta. \tag{4.4.10}$$

Let us recall that for any $f \in L^1(\Omega)$ we denote with f_Ω its average over Ω. Consider the function $f(x, \mathbf{y}, \mathbf{z}) = x^{-s}$, where $0 < s < \gamma k + 1$ will be chosen below. Observe that, under this condition on s, we have $f \in L^1(\Omega)$ and so f_Ω is well defined.

Now, for $\varepsilon > 0$ small, we define

$$f_\varepsilon(x, \mathbf{y}, \mathbf{z}) = \begin{cases} f(x, \mathbf{y}, \mathbf{z}) & \text{if } x > \varepsilon \\ \varepsilon^{-s} & \text{if } x \leq \varepsilon \end{cases}.$$

It follows from (4.4.10) that $f_\varepsilon \in L^p(\Omega, d_M^{p\beta})$.

Assume that the statement of Theorem 4.4 holds for some $\eta < \beta + \gamma - 1$. Then, for each ε, there exists $\mathbf{w}_\varepsilon = (w_{\varepsilon,1}, \mathbf{w}_{\varepsilon,n-1})$ such that div $\mathbf{w}_\varepsilon = f_\varepsilon - f_{\varepsilon,\Omega}$ verifying

$$\|\mathbf{Dw}_\varepsilon\|_{L^p(\Omega, d_M^{p\eta})} \leq C \|f_\varepsilon - f_{\varepsilon,\Omega}\|_{L^p(\Omega, d_M^{p\beta})}. \tag{4.4.11}$$

Recalling that $d_M \simeq x$, we have

$$\|f_\varepsilon - f_{\varepsilon,\Omega}\|^p_{L^p(\Omega, d_M^{p\beta})} \simeq \int_\Omega (f_\varepsilon - f_{\varepsilon,\Omega})^2 |f_\varepsilon - f_{\varepsilon,\Omega}|^{p-2} x^{p\beta} \, dx dy dz$$

$$= \int_\Omega \text{div } \mathbf{w}_\varepsilon (f_\varepsilon - f_{\varepsilon,\Omega}) |f_\varepsilon - f_{\varepsilon,\Omega}|^{p-2} x^{p\beta} \, dx dy dz,$$

and therefore, integrating by parts we obtain

$$\|f_\varepsilon - f_{\varepsilon,\Omega}\|^p_{L^p(\Omega, d_M^{p\beta})} \simeq - \int_\Omega w_{\varepsilon,1} \frac{\partial((f_\varepsilon - f_{\varepsilon,\Omega}) |f_\varepsilon - f_{\varepsilon,\Omega}|^{p-2} x^{p\beta})}{\partial x} \, dx dy dz. \tag{4.4.12}$$

To simplify notation let us define

$$h_\varepsilon := \frac{\partial((f_\varepsilon(x) - f_{\varepsilon,\Omega}) |f_\varepsilon(x) - f_{\varepsilon,\Omega}|^{p-2} x^{p\beta})}{\partial x}.$$

Recalling that $(x, \mathbf{y}, \mathbf{z}) = (x, y_1, \ldots, y_k, z_1, \ldots, z_{n-k-1})$, we can write $h_\varepsilon = \frac{\partial(y_1 h_\varepsilon)}{\partial y_1}$. Then, replacing in (4.4.12) and integrating by parts again, we obtain

$$\|f_\varepsilon - f_{\varepsilon,\Omega}\|^p_{L^p(\Omega, d_M^{p\beta})} \simeq \int_\Omega \frac{\partial w_{\varepsilon,1}}{\partial y_1} y_1 h_\varepsilon \, dx dy dz. \tag{4.4.13}$$

But, it is not difficult to see that, for $x < f_\Omega^{-\frac{1}{s}}/2$, we have

$$|h_\varepsilon| \leq Cx^{-s(p-1)+p\beta-1}.$$

Let us mention that we have considered x small enough to be away from the point where $f_\varepsilon(x) - f_{\varepsilon,\Omega} = 0$, because at that point h_ε blows up in the case $p < 2$.

Then, applying the Hölder inequality in (4.4.13), we obtain

$$\|f_\varepsilon - f_{\varepsilon,\Omega}\|^p_{L^p(\Omega, d_M^{p\beta})} \leq C\|\mathbf{D}w_\varepsilon\|_{L^p(\Omega, d_M^{p\eta})} \|y_1 x^{-s(p-1)+p\beta-1}\|_{L^q(\Omega, d_M^{-q\eta})}.$$

Therefore, using (4.4.11) and again $d_M \simeq x$, we conclude that

$$\|f_\varepsilon - f_{\varepsilon,\Omega}\|^p_{L^p(\Omega, d_M^{p\beta})} \leq C\|y_1 x^{-s(p-1)+p\beta-1} x^{-\eta}\|^q_{L^q(\Omega)}. \tag{4.4.14}$$

However, a straightforward computation shows that, a choice of a positive s in the range

$$\frac{\gamma k+1}{p} + \beta \leq s < \frac{\gamma k+1}{p} + \beta + (\beta + \gamma - \eta - 1)\frac{q}{p}, \tag{4.4.15}$$

leads to $\|f - f_\Omega\|_{L^p(\Omega, d_M^{p\beta})} = +\infty$ and $\|y_1 x^{-s(p-1)+p\beta-1} x^{-\eta}\|_{L^q(\Omega)} < \infty$. But, taking the limit $\varepsilon \to 0$ in (4.4.14) we get a contradiction. Since $\eta < \beta + \gamma - 1$ such an s exists. Recall that we also need $s < \gamma k + 1$, but, an s satisfying this restriction and (4.4.15) exists because $\frac{\gamma k+1}{p} + \beta < \gamma k + 1$. Indeed, this inequality follows from the assumption (4.4.15). Therefore, we conclude that a result as that in Theorem 4.4 is not valid under this relation on the exponents.

4.5 Korn$_p$ and div$_p$ for Linked Domains

In this section we show some techniques that allow to handle Korn$_p$ and div$_p$ in a complex domain Ω by dealing with more elementary parts of Ω. In this way the results of Sections 4.2 and 4.3 can be generalized to more general external cusps as we show in the next section.

Let us begin with some basic definitions useful along this section. For any domain Ω, with $\delta_M(\Omega)$ and $\delta_m(\Omega)$ we denote the outer and the inner diameters of Ω (i.e., the diameters of a minimal ball containing Ω and a maximal ball contained in Ω).

With $R \subset \mathbb{R}^n$ we denote a generic open rectangle $R \subset \mathbb{R}^n$ with edges *parallels to the coordinate axes*. The length of the R's i-th edge is $\ell_i(R)$ while for a cube Q we just use $\ell(Q)$. A pair of rectangles R_1, R_2 are called *C-comparable* and we write $R_1 \underset{C}{\sim} R_2$ if $\ell_i(R_1) \underset{C}{\sim} \ell_i(R_2)$ for $1 \leq i \leq n$. Sometimes we also write $R_1 \sim R_2$ in order to abbreviate notation. For a rectangle R, we denote its barycenter with c_R and with aR ($a > 0$), a dilated rectangle, centered in c_R, with edge lengths $\ell_i(aR) = a\ell_i(R)$. We say that R_1 and R_2 are touching rectangles if $R_1 \cap R_2 = \emptyset$ and $\overline{R}_1 \cap \overline{R}_2 = F$ with F a face of R_1 or R_2.

Finally, let us recall the notation $\cup \mathscr{C} := \cup_{S \in \mathscr{C}} S$, for a collection of sets $\mathscr{C}$. And that given two sets, we use $A \equiv B$ if they differ in measure zero.

Definition 4.5.1 *A (finite or countable) collection of rectangles* $\mathscr{C} = \{R_i\}$ *for which* $\sum_i |R_i| < \infty$, *is called* a chain of rectangles *if*

a) $\overline{R}_i \cap \overline{R}_j = \emptyset$ *for* $|i - j| > 1$,
b) *for any* i, R_i *and* R_{i+1} *are touching, and*
c) *there exists a* fixed *constant C such that* $R_i \underset{C}{\sim} R_{i+1}$, *for any* i.

The technique used below works for general chains of sets as long as they keep some basic properties of rectangles, so that we introduce the next definition.

Definition 4.5.2 (Quasi-Rectangles and Quasi-Cubes) *Let* $\mathscr{W} = \{\Omega_i\}$ *be a (finite or countable) collection of disjoint open sets. Assume that there exists a chain of rectangles* $\mathscr{C} = \{R_i\}$ *(in the sense of Definition 4.5.1) with* $R_i \subset \Omega_i \subset C_r R_i$ *for a fixed constant* C_r. *Then* $\mathscr{W} = \{\Omega_i\}$ *is called a chain of quasi-rectangles associated to the chain of rectangles* $\mathscr{C}$. *Each* Ω_i *is called a quasi-rectangle associated to* R_i. *Moreover, if each* R_i *is a cube, then each* Ω_i *is called a quasi-cube.*

Remark 4.7. Obviously a chain of rectangles is a chain of quasi-rectangles.

A chain of quasi-rectangles is a collection of disjoint sets. Using such a chain to build a domain requires to link the elements of the chain in some way.

Lemma 4.5. *Let* $\mathscr{W} = \{\Omega_i\}$ *be a chain of quasi-rectangles and* $\mathscr{C} = \{R_i\}$ *the chain of rectangles associated to* $\mathscr{W}$. *Then, there exist a family of* intermediate *rectangles* $\mathscr{C}_I = \{R_{i,i+1}\}$ *such that* $R_{i,i+1} \subset \overline{R}_i \cup \overline{R}_{i+1}$ *and*

1. $R_{i,i+1} \sim (R_{i,i+1} \cap R_i) \sim R_i \sim (R_{i,i+1} \cap R_{i+1}) \sim R_{i+1}$
2. $|R_{i,i+1}| \sim |(R_{i,i+1} \cap R_i)| \sim |R_i| \sim |(R_{i,i+1} \cap R_{i+1})| \sim |R_{i+1}|$
3. $|R_{i,i+1}| \sim |(R_{i,i+1} \cap \Omega_i)| \sim |\Omega_i| \sim |(R_{i,i+1} \cap \Omega_{i+1})| \sim |\Omega_{i+1}|$.

notice that the implicit constant in $\sim$ *can be taken the same for any* i.

Proof. Elementary using the definition of $\mathscr{W}$. $\square$

Definition 4.5.3 *Given a chain of quasi-rectangles* $\mathscr{W}$, *any collection of intermediate rectangles* $R_{i,i+1}$ *enjoying properties like those mentioned in Lemma 4.5 is denoted with* $\mathscr{C}_I = \{R_{i,i+1}\}$.

Definition 4.5.4 (Linked Domain) *Given a chain of quasi-rectangles* $\mathscr{W}$ *any* Ω *such that* $\Omega = \cup \{\mathscr{C}_I \cup \mathscr{W}\}$ *is a linked domain.*

Korn$_p$ and Poincaré: The first technique described below is based on rather simple ideas that also work for Poincaré inequalities: since the behavior of the constant in Korn's (resp. Poincaré) inequality *on rectangles* is known, we consider chains of rectangles, and use a discrete Hardy inequality to *pass* from one rectangle to another. As a result a weight arises naturally for the Korn (resp. Poincaré) inequality in the whole chain. Although the basic technique is rather standard for Poincaré inequalities, using, for instance, Whitney cubes instead of rectangles, in the elasticity framework this idea seems to go back to [44] in the context of nonlinear Korn type inequalities.

In order to match our presentation with the divergence counterpart introduced later we give a slightly modified version of that offered in [4].

Let us state the following discrete weighted inequality of Hardy type [70, page 52]:

Lemma 4.6. *Let* $\{u_i\}$ *and* $\{v_i\}$ *be sequences of non-negative weights and let* $1 < p \leq q < \infty$. *Then the inequality,*

$$\left[\sum_{j=1}^{\infty} u_j \left(\sum_{i=1}^{j} b_i\right)^q\right]^{\frac{1}{q}} \leq c \left[\sum_{j=1}^{\infty} v_j b_j^p\right]^{\frac{1}{p}}$$

holds for every non-negative sequences $\{b_i\}$ *if*

$$A = \sup_{k>0} \left(\sum_{j=k}^{\infty} u_j\right)^{\frac{1}{q}} \left(\sum_{j=0}^{k} v_j^{1-p'}\right)^{\frac{1}{p'}} < \infty.$$

The constant c *is* $c = M\mathbf{A}$, *where* M *depends only on* p *and* q.

From the previous Lemma we get.

Lemma 4.7. *Let* $\{r_i\}_i$ *and* $\mathbf{a} = \{a_i\}_i$ *be sequences such that* $\{r_i\}_i > 0$, *and* $\sum_i r_i = r < \infty$. *Let us denote*

$$\bar{a} = \frac{1}{r} \sum_j a_j r_j.$$

Then the inequality:

$$\left(\sum_{j=1}^{\infty} |a_j - \bar{a}|^p r_j\right)^{\frac{1}{p}} \leq c \left(\sum_{j=1}^{\infty} |a_{j+1} - a_j|^p r_{j+1}\right)^{\frac{1}{p}} \tag{4.5.1}$$

holds if

$$A = \sup_{k>0} \left(\sum_{j=k}^{\infty} r_j\right)^{\frac{1}{p}} \left(\sum_{j=0}^{k} r_j^{1-p'}\right)^{\frac{1}{p'}} < \infty \tag{4.5.2}$$

The constant c *is* $c = MA$ *where* M *depends only on* p.

Proof. Let us define the norm,

$$\|\mathbf{a}\|_p = \left(\sum_i |a_i|^p r_i\right)^{\frac{1}{p}}.$$

From Hölder's inequality, it holds $|\bar{a}|r \leq \|\mathbf{a}\|_p r^{\frac{1}{p'}}$ and then $\|\mathbf{a} - \bar{a}\|_p \leq 2\|\mathbf{a}\|_p$. Applying this last inequality with $\mathbf{a}$ replaced by $\mathbf{a} - a_0$, we obtain

$$\|\mathbf{a} - \bar{a}\|_p \leq 2\|\mathbf{a} - a_0\|_p.$$

Therefore:

$$\sum_i |a_i - \bar{a}|^p r_i \leq 2^p \sum_i |a_i - a_0|^p r_i \leq 2^p \sum_i \left(\sum_{j=1}^i |a_j - a_{j-1}| \right)^p r_i$$

And we conclude applying Lemma 4.6 with $u_i = v_i = r_i$, $q = p$ and $b_i = |a_i - a_{i-1}|$.
                                                                                                    □

Remark 4.8. Observe that if in Lemma 4.7, $\{r_i\}_{0 \leq i \leq N}$ *is finite and* $r_i \underset{C}{\sim} r$ *for any* i, then

$$A = \max_{N \geq k > 0} \left(\sum_{j=k}^N r_j \right)^{\frac{1}{p}} \left(\sum_{j=0}^k r_j^{1-p'} \right)^{\frac{1}{p'}} \leq CN \qquad (4.5.3)$$

since

$$A = \max_{N \geq k > 0} \left(\sum_{j=k}^N r_j \right)^{\frac{1}{p}} \left(\sum_{j=0}^k r_j^{1-p'} \right)^{\frac{1}{p'}} \leq C \max_{N \geq k > 0} \left(\sum_{j=k}^N r \right)^{\frac{1}{p}} \left(\sum_{j=0}^k r^{1-p'} \right)^{\frac{1}{p'}}$$

$$= C \max_{N \geq k > 0} \left((N - k + 1) r \right)^{\frac{1}{p}} \left((k+1) r^{1-p'} \right)^{\frac{1}{p'}} \leq C N^{\frac{1}{p} + \frac{1}{p'}} = CN$$

Remark 4.9. For any rectangle R (actually for any convex set) it is known (see, for instance, [13], Lemma 2.1, Pag. 4) that the constant in the Poincaré inequality can be bounded in terms of $\delta_M(R)$. In particular, for a chain $\mathscr{C} = \{R_i\}$ of *rectangles* each individual constant C_{P_i} is bounded by $C_{P_i} \leq C_P \delta_{M_i}$ being C_P an universal constant and $\delta_{M_i} = \delta_M(R_i)$. Similarly, $C_{div_{p_i}} \leq C_D \frac{\delta_{M_i}}{\delta_{m_i}}$, where $C_{div_{p_i}}$ is the constant for the div_p problem on R_i and $\delta_{m_i} = \delta_m(R_i)$, where C_D a fixed constant (see Section 2.5). Notice that this implies the same bound for $Korn_p$, that is: $C_{K_i} \leq C_K \frac{\delta_M(R_i)}{\delta_m(R_i)}$.

Remark 4.10. The estimate of the $Korn_p$ constant for a rectangle (see previous remark) is sharp. Take, for instance, $n = 2$, $0 < L_m \leq L_M$ and $\mathbf{u}(x,y) = (xy, -\frac{x^2}{2})$, defined over the rectangle $R = (0, L_M) \times (-\frac{L_m}{2}, \frac{L_m}{2})$ we have that

$$\|\mathbf{Du}\|^p_{L^p(R)^{n \times n}} \underset{C}{\sim} L_m L_M^{p+1} \quad \text{and} \quad \|\boldsymbol{\varepsilon}(\mathbf{u})\|^p_{L^p(R)^{n \times n}} \underset{C}{\sim} L_m^{p+1} L_M,$$

and therefore

$$\frac{\|\mathbf{Du}\|^p_{L^p(R)^{n \times n}}}{\|\boldsymbol{\varepsilon}(\mathbf{u})\|^p_{L^p(R)^{n \times n}}} \underset{C}{\sim} \left(\frac{L_M}{L_m} \right)^p.$$

Remark 4.11. Observe that if $\Omega_1 \subset \Omega_2$ and $|\Omega_2 \setminus \Omega_1| = 0$, then $K_{\Omega_1} \geq K_{\Omega_2}$ and $P_{\Omega_1} \geq P_{\Omega_2}$, where K_{Ω_j} and P_{Ω_j} are the constants for the second case of Korn's inequality and of Poincaré inequality on Ω_j, respectively. Consequently, the results that we prove for *linked* domains hold for any domain Ω such that $\cup(\mathscr{C}_I \cup \mathscr{W}) \subset \Omega$ and $\Omega \equiv \cup(\mathscr{C}_I \cup \mathscr{W})$.

Theorem 4.7 (Weighted *Korn$_p$* **for** *linked* **domains).** *Let Ω be a linked domain and $\mathscr{W} = \{\Omega_i\}$ the associated chain of quasi-rectangles. Let $1 < p < \infty$ and assume that the constant C_i, for Korn$_p$ on Ω_i can be bounded by $C_i \leq C_K \frac{\delta_{M_i}}{\delta_{m_i}}$ with $\delta_{M_i} := \delta_M(\Omega_i)$, $\delta_{m_i} := \delta_m(\Omega_i)$ and C_K independent on i. Then for any $\mathbf{u} \in W^{1,p}(\Omega)^n$ such that $\int_\Omega \mu(\mathbf{u}) = 0$ we have*

$$\|\mathbf{Du}\|_{L^p(\Omega)} \leq C(1+A)\|\boldsymbol{\varepsilon}(\mathbf{u})\|_{L^p(\Omega,\sigma^p)},$$

where A is defined in (4.5.2) with $r_j = |\Omega_j|$ and the weight σ is constant on each Ω_i with $\sigma|_{\Omega_i} \leq C \frac{\delta_{M_i}}{\delta_{m_i}}$.

   Moreover if the decay condition

$$|\Omega_{k+1}| \leq \alpha|\Omega_k| \quad \text{with } 0 < \alpha < 1. \tag{4.5.4}$$

holds, then

$$A \leq \left(\frac{1}{1-\alpha}\right)^{\frac{1}{p}}\left(\frac{1}{1-\alpha^{p'-1}}\right)^{\frac{1}{p'}}.$$

Proof. Let

$$A^i = \frac{1}{|\Omega_i|}\int_{\Omega_i}\mu(\mathbf{u}),$$

then

$$\|\mathbf{Du}\|^p_{L^p(\Omega)^{n\times n}} = \sum_i\|\mathbf{Du}\|^p_{L^p(\Omega_i)^{n\times n}} \leq \underbrace{C\sum_i\|\mathbf{Du} - A^i\|^p_{L^p(\Omega_i)^{n\times n}}}_{(I)}$$

$$+ \underbrace{C\sum_i\|A^i\|^p_{L^p(\Omega_i)^{n\times n}}}_{(II)}$$

(I) can be handled using that each Ω_i is a *Korn$_p$* domain with constant $C_i \leq C_K \frac{\delta_{M_i}}{\delta_{m_i}}$

$$I \leq C\sum_i C_i^p\|\boldsymbol{\varepsilon}(\mathbf{u})\|^p_{L^p(\Omega_i)^{n\times n}} \leq C\sum_i\|\boldsymbol{\varepsilon}(\mathbf{u})\|^p_{L^p(\Omega_i,\sigma^p)^{n\times n}} = C\|\boldsymbol{\varepsilon}(\mathbf{u})\|^p_{L^p(\Omega,\sigma^p)^{n\times n}}$$

For (II), apply inequality (4.5.1) with $r_j = |\Omega_j|$. Let us observe that $\sum|\Omega_i|A^i = 0$, therefore taking

$$A = \sup_{k>0}\left(\sum_{j\geq k}|\Omega_j|\right)^{\frac{1}{p}}\left(\sum_{j\leq k}|\Omega_j|^{1-p'}\right)^{\frac{1}{p'}}$$

we have

$$II = C \sum_i |A^i|^p |\Omega_i| \leq CA^p \sum_i |A^{i+1} - A^i|^p |\Omega_{i+1}|$$

where C is a constant depending only on n and p.

For each i, we consider now the intermediate rectangle $R_{i,i+1}$ of $\mathscr{C}_I$. Calling

$$A^{i,i+1} = \frac{1}{|R_{i,i+1}|} \int_{R_{i,i+1}} \mu(\mathbf{u})$$

we get, using extensively the properties given in Lemma 4.5,

$$II \leq CA^p \sum_i \left\{ |A^{i+1} - A^{i,i+1}|^p + |A^{i,i+1} - A^i|^p \right\} |\Omega_{i+1}|$$

$$\leq CA^p \sum_i \left\{ |A^{i+1} - A^{i,i+1}|^p |\Omega_{i+1} \cap R_{i,i+1}| + |A^{i,i+1} - A^i|^p |\Omega_i \cap R_{i,i+1}| \right\}$$

$$= CA^p \sum_i \left\{ \|A^{i+1} - A^{i,i+1}\|^p_{L^p(\Omega_{i+1} \cap R_{i,i+1})} + \|A^i - A^{i,i+1}\|^p_{L^p(\Omega_i \cap R_{i,i+1})} \right\}$$

$$\leq CA^p \sum_i \left\{ \|A^{i+1} - \mathbf{Du}\|^p_{L^p(\Omega_{i+1})} + \|\mathbf{Du} - A^{i,i+1}\|^p_{L^p(R_{i,i+1})} + \|\mathbf{Du} - A^i\|^p_{L^p(\Omega_i)} \right\}$$

Using the second case of Korn's inequality on each set, the fact that the respective constant can be bounded by $C_K \frac{\delta_{M_i}}{\delta_{m_i}}$, that $\Omega_{i+1} \cap \Omega_i = \emptyset$ while $R_{i,i+1}$ only touches Ω_i and Ω_{i+1} we have

$$II \leq CA^p \sum_i C_i^p \|\boldsymbol{\varepsilon}(\mathbf{u})\|^p_{L^p(\Omega_i)} \leq CA^p \|\boldsymbol{\varepsilon}(\mathbf{u})\|^p_{L^p(\Omega, \sigma^p)}$$

and then Korn's inequality follows using the bounds for (I) and (II).

To finish the proof assume (4.5.4). We get that $|\Omega_k| \leq \alpha^{k-i} |\Omega_i|$ for $0 \leq i \leq k$ and that $|\Omega_i| \leq \alpha^{i-k} |\Omega_k|$ for $i \geq k$, therefore

$$A = \sup_{k>0} \left(\sum_{j=k}^\infty |\Omega_j| \right)^{\frac{1}{p}} \left(\sum_{j=0}^k |\Omega_j|)^{1-p'} \right)^{\frac{1}{p'}} \leq |\Omega_k|^{\frac{1}{p}} \left(\sum_{j=0}^\infty \alpha^j \right)^{\frac{1}{p}} |\Omega_k|^{\frac{1}{p'}-1} \left(\sum_{j=0}^k \alpha^{(p'-1)j} \right)^{\frac{1}{p'}},$$

then

$$A \leq \left(\frac{1}{1-\alpha} \right)^{\frac{1}{p}} \left(\frac{1}{1-\alpha^{p'-1}} \right)^{\frac{1}{p'}}.$$

and the theorem follows. $\square$

The same technique used for Korn works for Poincaré inequality.

Theorem 4.8 (Weighted Poincaré inequality for *linked* **domains).** *Let Ω be a linked domain and $\mathscr{W} = \{\Omega_i\}$ the associated chain of quasi-rectangles. Let $1 < p < \infty$ and assume that the constant C_{P_i} for the Poincaré inequality on Ω_i can be bounded as $C_{P_i} \leq C_P \delta_{M_i}$. Then for any $u \in W^{1,p}(\Omega)$ we have*

$$\|u\|_{[L^p(\Omega)]} \le C(1+A)\|\nabla u\|_{L^p(\Omega,\sigma^p)},$$

where A is defined in (4.5.2) with $r_j = |\Omega_j|$ and the weight σ is constant on each Ω_i being $\sigma|_{\Omega_i} \le C\delta_{M_i}$.

Moreover if the decay condition (4.5.4) holds, then $A \le \left(\frac{1}{1-\alpha}\right)^{\frac{1}{p}}\left(\frac{1}{1-\alpha^{p'-1}}\right)^{\frac{1}{p'}}$.

Proof. Follows in the same fashion that the proof of Theorem 4.7. Here we only need to recall that the constant in Poincaré inequality for each intermediate rectangle $R_{i,i+1}$ (in general for a convex domain) depends only on the diameter of the rectangle that is comparable to δ_{M_i} due to the properties of $\mathscr{W}$.   □

Remark 4.12. Let us mention that Theorem 4.8 also holds for $p = 1$. The key point is that during the proof, Lemma 4.7 is invoked with the choice $r_i = |R_i|$ under the *decay condition* $|R_{i+1}| \le \alpha|R_i|$. In this case it is easy to see, by a direct proof of Lemma 4.6 for $p = 1$ taking $u_i = v_i = r_i$, that (4.5.1) holds for $p = 1$ and $c = \frac{2}{1-\alpha}$.

Theorem 4.9 (Weighted Korn$_p^u$ for linked domains). *Let Ω be a linked bounded domain and $\mathscr{W} = \{\Omega_i\}$ the associated chain of quasi-rectangles. Let $1 < p < \infty$ and assume that the constant C_i, for Korn$_p$ on Ω_i can be bounded by $C_i \le C_K\frac{\delta_{M_i}}{\delta_{m_i}}$ with $\delta_{M_i} := \delta_M(\Omega_i)$, $\delta_{m_i} := \delta_m(\Omega_i)$ and C_K independent on i. Assume the decay condition (4.5.4) and that $B \subset \Omega$ is a ball. Then for any $\mathbf{u} \in W^{1,p}(\Omega)^n$*

$$\|\mathbf{Du}\|_{L^p(\Omega)^{n\times n}} \le C\left\{\|\boldsymbol{\varepsilon}(\mathbf{u})\|_{L^p(\Omega,\sigma^p)^{n\times n}} + \|\mathbf{u}\|_{[L^p(B)^n]}\right\}, \tag{4.5.5}$$

where the weight σ is constant on each element of $\mathscr{W}$ and can be bounded as $\sigma|_{\Omega_i} \le C\frac{\delta_{M_i}}{\delta_{m_i}}$.

Proof. The weighted Korn$_{p,B}^u$ follows from weighted Korn$_p$ (Theorem 4.7) in exactly the same way deployed in Theorem 3.1 for the unweighted inequalities.   □

The reader may wonder how general a family of quasi-rectangles could be in order to satisfy the requirement of previous theorems and in particular the conditions $C_{K_i} \le C_K\frac{\delta_{M_i}}{\delta_{m_i}}$ and $C_{P_i} \le C_P\delta_{M_i}$ (that hold, for instance, for convex quasi-rectangles). In order to give a hint we exploit recursively the concept of chain: consider a bounded John domain B and call C_K its Korn$_p$ constant and C_P its Poincaré constant. Take a central cube Q such that $Q \subset B \subset C_Q Q$. Since B is a fixed domain and the Korn$_p$ constant is invariant by scalings (while the Poincaré constant scales with $\delta_M(B)$) we can easily find finite chains of quasi-cubes providing quasi-rectangles satisfying our requirements. Moreover, as we see below, there is no need to have the same constant for every element of the chain but a uniform bound. In Figure 4.3 we show a nontrivial quasi-rectangle.

Corollary 4.4. *Let $\mathscr{W} = \{\Omega_i\}_{1 \le i \le N}$ be a finite chain of quasi-cubes associated to a finite chain of cubes $\mathscr{C} = \{Q_i\}_{1 \le i \le N}$ with their centers placed along a straight line parallel to an axis. Assume that for $1 \le i \le N$, $\ell(Q_i) = \ell$ and that $C_{K_i} \le C_K$ and $C_{P_i} \le C_P \delta_M(\Omega_i)$. Then, any linked domain Ω associated to $\mathscr{W}$ is a quasi-rectangle associated to R, being R the minimal rectangle containing the chain $\mathscr{C}$ and the constants for Korn$_p$ and Poincaré of Ω can be bounded by $C_K \frac{\delta_M(\Omega)}{\delta_m(\Omega)}$ and $C_P \delta_M(\Omega)$ respectively.*

Proof. We show only the bound for *Korn$_p$* since the other one follows exactly in the same way. First notice that by definition of chain of quasi-cubes $\frac{\delta_M(\Omega_i)}{\delta_m(\Omega_i)} \sim \frac{\delta_M(Q_i)}{\delta_m(Q_i)} \sim 1$ and therefore we can use Theorem 4.7 applied to the *finite* chain $\mathscr{W}$. Observe that for the resulting weight we have $\sigma|_{\Omega_i} \le C_K$ and therefore σ can be written outside the norm using C_K. On the other hand, instead of the decay condition (useful for infinite chains) we notice that $|\Omega_i| \underset{C}{\sim} |Q_i| = \ell^n$ and use Remark 4.8 to get $A \le CN$. Finally, $N = \frac{N\ell}{\ell} \le \frac{\delta_M(R)}{\delta_m(R)} \le C \frac{\delta_M(\Omega)}{\delta_m(\Omega)}$.   □

Observe that this corollary only provides quasi-rectangles with interior rectangles having $n-1$ equal short edges and a long one.

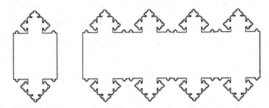

Fig. 4.3 Left: a John domain providing a quasi-cube and Right: a quasi-rectangle obtained as a linked domain of finite quasi-cubes. In the latter, the constant for the second case of Korn's inequality scales linearly with the number of quasi-cubes

Similar ideas can be used to handle the divergence operator as we see below.

div$_p$ problem: There is a simple idea already exploited by Bogovskiĭ. Assume that $\Omega = \Omega_0 \cup \Omega_1$ is a domain, where each Ω_i is a subdomain. Let $\{\phi_0, \phi_1\}$ be a partition of the unity subordinated to $\{\Omega_0, \Omega_1\}$.

For $f \in L_0^p(\Omega)$, we can write

$$f = f_0 + f_1 = f\phi_0 + f\phi_1$$

which gives a decomposition of f in terms of functions f_i locally supported in each Ω_i. In this decomposition, however, the vanishing average property is lost at the level of Ω_i. In order to retain this property the following procedure can be applied.

Define $B_1 = \Omega_1 \cap \Omega_0$ and write

$$f = \left(f_0 + \frac{\chi_{B_1}}{|B_1|} \int_{\Omega_1} f_1 \right) + \left(f_1 - \frac{\chi_{B_1}}{|B_1|} \int_{\Omega_1} f_1 \right).$$

The gaining obtained so far is that each function between parentheses is still supported in Ω_i and with zero average (we use that $f \in L_0^p(\Omega)$). It is easy to see that if $\Omega = \cup_{0 \le i \le 2}\{\Omega_i\}$ we can take $B_i = \Omega_i \cap \Omega_{i-1}$

$$\tilde{f}_0 = -\frac{\chi_{B_1}}{|B_1|}\int_{\Omega_1 \cup \Omega_2} f_1 + f_2,$$

$$\tilde{f}_1 = \frac{\chi_{B_1}}{|B_1|}\int_{\Omega_1 \cup \Omega_2} f_1 + f_2 - \frac{\chi_{B_2}}{|B_2|}\int_{\Omega_2} f_2,$$

$$\tilde{f}_2 = \frac{\chi_{B_2}}{|B_2|}\int_{\Omega_2} f_2.$$

and with this choice $f = \sum_{0 \le i \le 2}(f_i - \tilde{f}_i)$, each $(f_i - \tilde{f}_i)$ is supported in Ω_i and has vanishing average (the vanishing mean value of f guarantees that $f_0 - \tilde{f}_0$ also integrates zero).

This trick can be extended to a rather general collections of subdomains (actually *trees* of subdomains as originally treated in [75] where the reader can find more details with the warning that weighted spaces are not defined in the same way). Below we restrict our attention to linked domains in order to match with the previous presentation devoted to Korn$_p$. For this reason, we only consider here a very restricted case of *trees*.

Definition 4.1. Let $\tilde{\mathscr{W}} = \{\tilde{\Omega}_i\}$ be a family of finite or countable domains for which $\tilde{\Omega}_i \cap \tilde{\Omega}_j = \emptyset$ only if $|i - j| > 1$. We say that $\tilde{\mathscr{W}}$ is an *overlapping* chain of domains.

Notice that the overlapping in such a chain $\tilde{\mathscr{W}}$ is a finite (actually it is bounded by 3). Finally, let us introduce the following handy notation, for any integer number $1 \le k$, B_k denotes the intersection $B_k := \tilde{\Omega}_k \cap \tilde{\Omega}_{k-1}$ and W_k the decreasing family of subdomains given by $W_k := \cup\{\tilde{\Omega}_i\}_{i \ge k} \subset \Omega := \cup\tilde{\mathscr{W}}$.

Following previous considerations, let $\{\phi_i\}$ be a partition of the unity subordinated to $\{\tilde{\Omega}_i\}$ and write $f = \sum f_i$, where $f_i = f\phi_i$. We have

$$\sum \|f_i\|_{L^p(\tilde{\Omega}_i)}^p \le C\|f\|_{L^p(\Omega)}^p,$$

where the constant C comes from the overlapping of the sets belonging to $\tilde{\mathscr{W}}$ and can be taken equal to 3. Define now $\tilde{f}_i$ as

$$\tilde{f}_i(x) := \frac{\chi_i(x)}{|B_i|}\int_{W_i}\sum_{k \ge i} f_k - \frac{\chi_{i+1}(x)}{|B_{i+1}|}\int_{W_{i+1}}\sum_{k \ge i+1} f_k, \tag{4.5.6}$$

where χ_i is the characteristic function of B_i. In the particular case when $i = 0$, the formula (4.5.6) gives

$$\tilde{f}_0(x) = -\sum \frac{\chi_1(x)}{|B_1|}\int_{W_s}\sum_{k \ge 1} f_k,$$

since $f \in L_0^p(\Omega)$.

Let us introduce the operator $T : L^p(\Omega) \to L^p(\Omega)$ defined by

$$Tf(x) := \sum_{i \geq 1} \frac{\chi_i(x)}{|W_i|} \int_{W_i} |f|. \qquad (4.5.7)$$

In the next theorem it is proved that f can be written as $\sum_{t \in \Gamma} f_t - \tilde{f}_t$, although first we deal with the continuity of the operator T.

First let us recall the following

Definition 4.2. Let T be an operator mapping measurable functions on measurable functions. If for any $\lambda > 0$,

$$|\{x : |Tf(x)| > \lambda\}| \leq \frac{C}{\lambda} \|f\|_{L^1(\Omega)},$$

then T is called of weak-type[1] $(1,1)$.

Lemma 4.8. *The operator T defined in (4.5.7) is of weak-type $(1,1)$ and continuous as an operator from $L^p(\Omega)$ into $L^p(\Omega)$ when $1 < p \leq \infty$. Moreover, the continuity constant C can be bounded by $2 \left(\dfrac{3p}{p-1} \right)^{1/p}$.*

Proof. We prove first the extremal cases showing the continuity for $p = \infty$ and the weak-type $(1,1)$. Then, finally, using Marcinkiewicz interpolation (see Theorem 2.4 on [34]) we extend the result to all $1 < p < \infty$.

T is an average of f when it is not zero, thus by a straightforward calculation, it can be proved that T is continuous from L^∞ to L^∞, with norm $\|T\|_{L^\infty \to L^\infty} \leq 1$. In order to prove the weak $(1,1)$ continuity, and given $\lambda > 0$, we pick the first index i_0 such that $\frac{1}{|W_{i_0}|} \int_{W_{i_0}} |f| > \lambda$, then we have

$$|\{x \in \Omega : Tf(x) > \lambda\}| \leq |W_{i_0}|$$
$$< \frac{1}{\lambda} \int_{W_{i_0}} |f| \leq \frac{3}{\lambda} \|f\|_{L^1(\Omega)},$$

where again 3 controls the overlapping of the collection $\mathscr{W}$. Thus, T is of weak-type $(1,1)$ with norm bounded by 3.

As a consequence, from standard interpolation arguments, $T : L^p(\Omega) \to L^p(\Omega)$ is continuous with norm bounded by $2 \left(\dfrac{3p}{p-1} \right)^{1/p}$. $\qquad \square$

---

[1] Weak-type is weaker than continuity as an operator $T : L^1(\Omega) \to L^1(\Omega)$ (usually called strong-type $(1,1)$ in Harmonic Analysis [84]). Indeed, for any $\lambda > 0$, the inequality

$$\lambda |\{x : |Tf(x)| > \lambda\}| \leq \|Tf\|_{L^1(\Omega)},$$

is immediate.

Now, we define a piecewise constant weight $\omega : \Omega \to \mathbb{R}_+$ by

$$
\omega(x) := \begin{cases} \dfrac{|B_i|}{|W_i|} & \text{if } x \in B_i \text{ for some } 1 \leq i \\[2ex] c_i & \text{if } x \in \tilde{\Omega}_i \setminus B_i \cup B_{i-1} \text{ for some } 1 \leq i. \end{cases}
\tag{4.5.8}
$$

the constant $0 < c_i \leq 1$ is arbitrary at this point (in [75], is taken $c_i = 1$) and will be appropriately chosen later on. By definition of $\mathscr{W}$, the weight is well defined and $0 < \omega(x) \leq 1$ for all $x \in \Omega$.

Theorem 4.10. *Let $\mathscr{W}$ be a chain of overlapping domains and consider $\Omega = \cup \mathscr{W}$. Given $f \in L^1(\Omega)$, and $1 < p < \infty$, the decomposition $f = \sum_{0 \leq i} f_i - \tilde{f}_i$ defined on (4.5.6) satisfies that $Supp(f_i - \tilde{f}_i) \subset \tilde{\Omega}_i$, $\int_{\tilde{\Omega}_i} f_i - \tilde{f}_i = 0$ for all $1 \leq i$, $\int_{\tilde{\Omega}_0} f_0 - \tilde{f}_0 = \int_\Omega f$, and*

$$
\sum_{0 \leq i} \| f_i - \tilde{f}_i \|^p_{L^p(\tilde{\Omega}_i, \omega^p)} \leq C \| f \|^p_{L^p(\Omega)},
\tag{4.5.9}
$$

where the constant C can be bounded by $2^p 3 \left(1 + \dfrac{2^{p+1} p}{p-1} \right)$.

Proof. By construction $Supp f_i \subset \tilde{\Omega}_i$ while $B_i, B_{i+1} \subset \tilde{\Omega}_i$, therefore $Supp(f_i - \tilde{f}_i) \subset \tilde{\Omega}_i$.

Obviously $\sum_{i=0}^{\infty} \tilde{f}_i(x) = 0$ if $x \notin \cup\{B_i\}$. On the other hand, if $x \in B_j$, then

$$
\sum_{i=0}^{\infty} \tilde{f}_i(x) = \tilde{f}_{j-1}(x) + \tilde{f}_j(x) = 0,
$$

since $B_{j-1} \cap B_j = \emptyset = B_j \cap B_{j+1}$ and therefore $\tilde{f}_{j-1}(x) = -\dfrac{1}{|B_j|} \int_{W_j} \sum_{k \geq j} f_k = -\tilde{f}_j(x)$.

As a consequence,

$$
\sum_{0 \leq i} f_i(x) - \tilde{f}_i(x) = \sum_{0 \leq i} f_i(x) - \sum_{0 \leq i} \tilde{f}_i(x) = f(x) + 0.
$$

On the other hand, from (4.5.6) we readily get

$$
\int_{\tilde{\Omega}_i} \tilde{f}_i = \int_{W_i} \sum_{i \leq k} f_k - \int_{W_{i-1}} \sum_{i+1 \leq k} f_k = \int_{W_i} f_i = \int_{\tilde{\Omega}_i} f_i,
$$

where we have used that $Supp \left(\sum_{i+1 \leq k} f_k \right) \subset W_{i+1}$ for the second equality and that $Supp f_i \subset \tilde{\Omega}_i \subset W_i$ for the third one. As a consequence we have $\int_{\tilde{\Omega}_i} f_i - \tilde{f}_i = 0$, while a straightforward calculation shows that $\int_{\tilde{\Omega}_0} f_0 - \tilde{f}_0 = \int_\Omega f$.

In order to show (4.5.9), we write

$$
\sum_{0 \leq i} \| f_i - \tilde{f}_i \|^p_{L^p(\tilde{\Omega}_i, \omega^p)} \leq 2^p \sum_{0 \leq i} \| f_i \|^p_{L^p(\tilde{\Omega}_i, \omega^p)} + 2^p \sum_{0 \leq i} \| \tilde{f}_i \|^p_{L^p(\tilde{\Omega}_i, \omega^p)}
$$

and notice that the first term is easy to bound since $0 \le \phi_i$, $\omega \le 1$ and the finite overlapping of $\mathscr{W}$

$$2^p \sum_{0 \le i} \|f_i\|^p_{L^p(\tilde{\Omega}_i, \omega^p)} \le 2^p 3 \|f\|^p_{L^p(\Omega)}.$$

For the second term, since $B_i \cap B_{i+1} = \emptyset$, we have for $1 \le i$

$$
\begin{aligned}
|\tilde{f}_i(x)|^p \omega(x)^p &\le \left(\frac{\omega(x)\chi_i(x)}{|B_i|} \int_{W_i} |f| + \frac{\omega(x)\chi_{j+1}(x)}{|B_{j+1}|} \int_{W_{j+1}} |f| \right)^p \\
&= \left(\frac{\chi_i(x)}{|W_i|} \int_{W_i} |f| + \frac{\chi_{i+1}(x)}{|W_{i+1}|} \int_{W_{i+1}} |f| \right)^p \\
&= \left(\frac{\chi_i(x)}{|W_i|} \int_{W_i} |f| \right)^p + \left(\frac{\chi_{i+1}(x)}{|W_{i+1}|} \int_{W_{i+1}} |f| \right)^p,
\end{aligned}
$$

with a similar result for $i = 0$. Hence,

$$
\begin{aligned}
\sum_{0 \le i} \int_{\tilde{\Omega}_i} |\tilde{f}_i(x)|^p \omega(x)^p &\le 2 \int_{\Omega} \sum_{i \le 1} \left(\frac{\chi_i(x)}{|W_i|} \int_{W_i} |f| \right)^p \\
&= 2 \int_{\Omega} \left(\sum_{1 \le i} \frac{\chi_i(x)}{|W_i|} \int_{W_i} |f| \right)^p.
\end{aligned}
$$

therefore

$$2^p \sum_{0 \le i} \|\tilde{f}_i\|^p_{L^p(\tilde{\Omega}_i, \omega^p)} \le 2^{p+1} \int_{\Omega} Tf(x)^p \le 2^{2p+1} \frac{3p}{p-1} \|f\|^p_{L^p(\Omega)},$$

thanks to Lemma 4.8 and as a consequence (4.5.9) follows. $\square$

Now we apply Theorem 4.10 to show the existence of a weighted solution of the divergence problem on linked domains. In order to do that we need to write such a domain using an appropriate overlapping chain of domains.

Remark 4.13. For a linked domain Ω we consider the associated chain of quasi-rectangles $\mathscr{W} = \{\Omega_i\}$ and intermediate rectangles $\mathscr{C}_I = \{R_{i,i+1}\}$. Define for any $0 \le i$, $\tilde{\Omega}_{2i} = \Omega_i$ and $\tilde{\Omega}_{2i+1} = R_{i,i+1}$. The collection $\tilde{\mathscr{W}} = \{\tilde{\Omega}_i\}$ is an overlapping chain of domains with overlapping number 3. In this particular case we say also that $\tilde{\mathscr{W}} = \{\tilde{\Omega}_i\}$ is an *overlapping chain of quasi-rectangles*.

Before stating the next result we pick c_i in (4.5.8), still arbitrarily as long as we guarantee a bounded oscillation of ω in each $\tilde{\Omega}_i$, that is

$$\min_{\tilde{\Omega}_i} \omega \le \max_{\tilde{\Omega}_i} \omega \le C \min_{\tilde{\Omega}_i} \omega. \qquad (4.5.10)$$

This is possible thanks to the properties of our chains. Indeed, on the one hand,

$$\frac{|\tilde{\Omega}_{i-1}|}{|\tilde{\Omega}_i|} \frac{1}{(1+C)} \le \frac{|\tilde{\Omega}_{i-1}||W_i|}{|\tilde{\Omega}_i||W_{i-1}|} \le \frac{|\tilde{\Omega}_{i-1}|}{|\tilde{\Omega}_i|}$$

since for an overlapping chain of quasi-rectangles $\frac{1}{C}|\tilde{\Omega}_i| \leq |\tilde{\Omega}_{i-1}| \leq C|\tilde{\Omega}_i|$. Moreover (see Lemma 4.5), $|B_i| \leq |\tilde{\Omega}_i| \leq C|B_i|$, and therefore

$$\frac{1}{C} \leq \frac{\frac{|B_{i-1}|}{|W_{i-1}|}}{\frac{|B_i|}{|W_i|}} \leq C. \tag{4.5.11}$$

Thanks to (4.5.11) we can take, for instance, $c_i = \left(\frac{|B_i|}{|W_i|} + \frac{|B_{i-1}|}{|W_{i-1}|} \right)/2$ satisfying (4.5.10).

This fact allows to introduce a simplified weight. Observe that for a linked domain we can write $\Omega \equiv \cup \{\tilde{\Omega}_{2i}\}$ then define

$$\bar{\omega}(x) := \frac{|B_{2i}|}{|W_{2i}|} \quad \text{if} \quad x \in \tilde{\Omega}_{2i}. \tag{4.5.12}$$

Notice that the weight is well defined (almost everywhere in Ω) since $\tilde{\Omega}_{2i} \cap \tilde{\Omega}_{2j} = \Omega_i \cap \Omega_j = \emptyset$ if $i \neq j$ and that thanks to (4.5.10)

$$\omega(x) \underset{C}{\sim} \bar{\omega}(x) \tag{4.5.13}$$

Theorem 4.11. *Let $\Omega = \cup \tilde{\mathcal{W}}$ be a linked domain with $\{\tilde{\mathcal{W}}\}$ an associated chain of overlapping quasi-rectangles. Assume that the constant C_{D_i} for the div$_p$ problem on $\tilde{\Omega}_i$ can be bounded by $C_{D_i} \leq C_D \frac{\delta_{M_i}}{\delta_{m_i}}$. Then, for any $1 < p < \infty$ and $f \in L_0^p(\Omega)$ there exists a solution $\mathbf{u} \in W_0^{1,p}(\Omega, (\frac{\bar{\omega}}{\sigma})^p)^n$ of div $\mathbf{u} = f$ such that*

$$\|\mathbf{Du}\|_{L^p(\Omega,(\frac{\bar{\omega}}{\sigma})^p)} \leq C\|f\|_{L^p(\Omega)}, \tag{4.5.14}$$

where σ can be taken piecewise constant in each Ω_i and $\sigma|_{\Omega_i} = \frac{\delta_{M_i}}{\delta_{m_i}}$.

Proof. Using Theorem 4.10 we can decompose the integrable function f as

$$f = \sum_{0 \leq i} f_i - \tilde{f}_i,$$

where $f_i - \tilde{f}_i \in L^p(\tilde{\Omega}_i)$, with vanishing mean value, and

$$\sum_{0 \leq i} \|f_i - \tilde{f}_i\|_{L^p(\tilde{\Omega}_i, \omega^p)}^p \leq C\|f\|_{L^p(\Omega)}^p.$$

By hypothesis, on each $\tilde{\Omega}_i$ it is possible to solve div$_p$ by means of a function $\mathbf{u}_i \in W_0^{1,p}(\tilde{\Omega}_i)$ such that

$$\frac{\delta_{m_i}}{\delta_{M_i}} \|\mathbf{Du}_i\|_{L^p(\tilde{\Omega}_i)} \leq C\|f_i - \tilde{f}_i\|_{L^p(\tilde{\Omega}_i)}, \tag{4.5.15}$$

therefore, writing $\mathbf{u} := \sum_{i\geq 0} \mathbf{u}_i$, using that $\Omega \equiv \cup_{i\geq 0} \{\tilde{\Omega}_{2i}\}$, the finite overlapping number of $\tilde{\mathscr{W}}$ and the fact that $\frac{\delta_{m_{i-1}}}{\delta_{M_{i-1}}} \underset{C}{\sim} \frac{\delta_{m_i}}{\delta_{M_i}}$ we get

$$\|\mathbf{Du}\|_{L^p(\Omega,(\frac{\omega}{\sigma})^p)}^p \leq C \sum_{i\geq 0} \left(\frac{\delta_{m_{2i}}}{\delta_{M_{2i}}} \frac{|B_{2i}|}{|W_{2i}|} \right)^p \|\mathbf{Du}\|_{L^p(\tilde{\Omega}_{2i})}^p \leq C \sum_{i\geq 0} \left(\frac{\delta_{m_i}}{\delta_{M_i}} \frac{|B_i|}{|W_i|} \right)^p \|\mathbf{Du}_i\|_{L^p(\tilde{\Omega}_i)}^p.$$

taking this into account, using (4.5.15) and (4.5.13), yields

$$\|\mathbf{Du}\|_{L^p(\Omega,(\frac{\omega}{\sigma})^p)}^p \leq C \sum_{i\geq 0} \left(\frac{|B_i|}{|W_i|} \right)^p \|f_i - \tilde{f}_i\|_{L^p(\tilde{\Omega}_i)}^p \leq C \sum_{i\geq 0} \|f_i - \tilde{f}_i\|_{L^p(\tilde{\Omega}_i,\omega)}^p \leq C \|f\|_{L^p(\Omega)}^p,$$

proving that $\mathbf{u}$ belongs to $W^{1,p}(\Omega,(\frac{\omega}{\sigma})^p)^n$ and the estimate (4.5.14). On the other hand, we can easily see that $\mathbf{u} \in W_0^{1,p}(\Omega,(\frac{\omega}{\sigma})^p)^n$, since for each finite sum $\sum_{0\leq i\leq N} \mathbf{u}_i \in W_0^{1,p}(\Omega,(\frac{\omega}{\sigma})^p)^n$ while $\sum_{i\geq N+1} \|\mathbf{Du}_i\|_{L^p(\tilde{\Omega}_i,(\frac{\omega}{\sigma})^p)}^p$ can be taken arbitrarily small for a large N due to the fact that $\sum_{i\geq 0} \|\mathbf{Du}_i\|_{L^p(\tilde{\Omega}_i,(\frac{\omega}{\sigma})^p)}^p < \infty$.   □

As an elementary application of the previous result we can derive estimates for quasi-rectangles. In some way we can mimic the procedure used in Corollary 4.4 applying recursively Theorem 4.11.

Corollary 4.5.1 *Let $\mathscr{W} = \{\Omega_i\}_{1\leq i\leq N}$ be a finite chain of quasi-cubes associated to a finite chain of cubes $\mathscr{C} = \{Q_i\}_{1\leq i\leq N}$ with their centers placed along a straight line parallel to an axis. Assume that for $1 \leq i \leq N$, $\ell(Q_i) = \ell$ and that each individual constant for the div_p problem verifies $C_{D_i} \leq C_D$. Then any linked domain Ω associated to $\mathscr{W}$ is a quasi-rectangle associated to R, being R the minimal rectangle containing the chain $\mathscr{C}$ and the constant for the div_p problem on Ω can be bounded by $C_D \frac{\delta_M(\Omega)}{\delta_m(\Omega)}$.*

Proof. Arguing as in Corollary 4.4 we observe that $\frac{\delta_M(\Omega_i)}{\delta_m(\Omega_i)} \sim \frac{\delta_M(Q_i)}{\delta_m(Q_i)} \sim 1$, hence σ^{-1} can be written as a constant outside the norm. On the other hand, we obviously have $\min_{1\leq i\leq N} \frac{|B_i|}{|W_i|} \geq C\frac{1}{N}$ and the Corollary follows since $N \sim \frac{\delta_M(\Omega)}{\delta_m(\Omega)}$.   □

Let us mention that Theorems 4.7, 4.9, and 4.11 can be straightforwardly generalized to versions where part of the weight is moved between the norms on the left- and right-hand sides. Moreover, a moment of reflection shows that it is possible to add arbitrary weights on both sides if they have bounded oscillation over each individual (and neighboring) quasi-rectangle. For these cases we refer the reader to [4, 75].

4.6 From Linked Domains to More General External Cusps

Now we consider *regular* external cusps, that is, with an isolated singular point. However, in this case we allow the rest of the boundary to be less regular than in (4.2.1). To simplify notation we assume that the singular point is placed at 0 and introduce a function φ, *that does not need to depict the precise profile of the domain* but only to give a qualitative description of its narrowing towards 0.

Within this context we need to have a notion of an axis as a segment to which the cusp becomes tangential. We assume this to be the x_1 axis denoted with $\hat{x}_1$. We restrict our attention to rectangles with $n-1$ short edges of equal size and a long edge (the one along $\hat{x}_1$). Usual external cusps are easily described using a "profile" function φ, as presented in (4.2.1) with $\varphi(x_1) \underset{C}{\sim} x_1^\gamma$. For that reason we consider the following definition.

Definition 4.6.1 *Any nondecreasing C^1 function $\varphi : \mathbb{R}_{\geq 0} \longrightarrow \mathbb{R}_{\geq 0}$ such that φ' is nondecreasing, $\varphi(0) = \varphi'(0) = 0$ and for any $0 < k$ there exists a constant C such that $\frac{\varphi(x)}{\varphi(kx)} \underset{C}{\sim} 1 \underset{C}{\sim} \frac{\varphi'(x)}{\varphi'(kx)}$ is called a* profile function.

Remark 4.14. For any $\gamma \geq 1$ the function x^γ is a profile function.

Definition 4.6.2 *Any sequence $\{a_i\}$ of strictly decreasing positive numbers such that*

1. $a_i \to 0$ *for* $i \to \infty$,
2. $a_{i+1} < a_i \leq Ca_{i+1}$,
3. $(a_i - a_{i+1}) \leq \alpha(a_{i-1} - a_i)$,

with $0 < \alpha < 1$ and $0 < C$ fixed constants, is called a profile sequence.

Now we introduce a more general class of external cusps.

Definition 4.6.3 (Generalized External Cusp) *Let $\{a_i\}$ and φ be a profile sequence and a profile function, respectively. Let $\mathcal{W} = \{\Omega_i\}$ be a chain of quasi-rectangles with an associated chain of rectangles $\mathcal{C} = \{R_i\}$ for which*

$$R_i := \{\mathbf{x} = (x, \mathbf{y}) \in \mathbb{R} \times \mathbb{R}^{n-1} : a_{i+1} < x_1 < a_i \, \|\mathbf{y} - \mathbf{c}_{\mathbf{R}_i}^{\mathbf{y}}\|_\infty < \varphi(a_i)\}, \qquad (4.6.1)$$

being $(c_{R_i}^x, \mathbf{c}_{\mathbf{R}_i}^{\mathbf{y}})$ the barycenter c_{R_i} of R_i and $\|\cdot\|_\infty$ the infinite norm of vectors. Then, any linked domain Ω associated to $\mathcal{W}$ is called a generalized external cusp.

The theory presented along the previous section applies straightforwardly to *generalized external cusps*. In Figure 4.4 we show some elementary examples of generalized external cusps. We take $\{\frac{1}{2^i}\}$ and $\varphi(x) = x^2$ as a profile sequence and function, respectively. The left figure is just an external cusp built with a chain of rectangles obeying (4.6.1), in the center of the figure an external cusp with a locally smooth boundary away from the origin. In this case the interior chain of rectangles is like the one on the left but not perfectly aligned. On the right the same kind of cusp with a boundary with poor regularity is exhibited.

Let us notice that for any generalized external cusp we have

$$\frac{\delta_m(\Omega_i)}{\delta_M(\Omega_i)} \geq C\frac{\varphi(a_i)}{a_{i-1} - a_i} \geq C\frac{\varphi(a_{i-1})}{a_{i-1} - a_i} \geq C\frac{\varphi(a_{i-1}) - \varphi(a_i)}{a_{i-1} - a_i} \geq C\varphi'(a_i). \qquad (4.6.2)$$

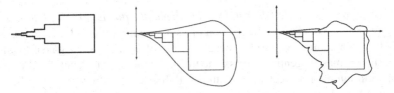

Fig. 4.4 Simple examples of generalized external cusps as linked domains. Left: a linked domain given by a chain of perfectly aligned rectangles, Center: external cusp with smooth boundary showing its central chain of rectangles, Right: external cusp with poor regularity and the same central chain of rectangles

This is a key inequality in the following results where in each case we recall that involved quasi-rectangles can have a very intricate boundary (for instance, such as those showed in Figure 4.3). First we show that Corollary 4.3 can be generalized.

Corollary 4.5. *Let Ω be a generalized external cusp and $\mathscr{W} = \{\Omega_i\}$ the associated chain of quasi-rectangles. Let $1 < p < \infty$ and assume that the constant C_i, for Korn$_p$ on Ω_i can be bounded by $C_i \leq C_K \frac{\delta_{M_i}}{\delta_{m_i}}$ with $\delta_{M_i} := \delta_M(\Omega_i)$, $\delta_{m_i} := \delta_m(\Omega_i)$ and C_K independent on i. For a fixed ball $B \subset \Omega$ and any $\mathbf{u} \in W^{1,p}(\Omega)^n$ it holds*

$$\|D\mathbf{u}\|_{L^p(\Omega)^{n \times n}} \leq C\left\{ \|\boldsymbol{\varepsilon}(\mathbf{u})\|_{L^p(\Omega,\sigma^p)^{n \times n}} + \|\mathbf{u}\|_{[L^p(B)^n]} \right\}, \qquad (4.6.3)$$

where the weight $\sigma(\mathbf{x})$ depends only on the first variable and can be defined as $\sigma(x,y) := \frac{1}{\varphi'(x)}$.

Proof. In order to use Theorem 4.9 we need to show the decay condition (4.5.4) and it follows straightforwardly from (3) in Definition 4.6.2 and the fact that φ is nondecreasing. The corollary follows using (4.6.2). $\qquad\square$

In the same fashion we can generalize Theorem 4.4. Notice that in this theorem for $M = \{\mathbf{0}\}$, $\beta = 0$ and $\eta = \gamma - 1$ we can write, using Remark 4.3, the following

$$\|D\mathbf{u}\|_{W^{1,p}(\Omega,x^{p(\gamma-1)})^n} \leq C\|f\|_{L^p(\Omega}. \qquad (4.6.4)$$

Now we show how this result looks like for generalized external cusps.

Corollary 4.6. *Let Ω be a generalized external cusp $\{\tilde{\mathscr{W}}\}$ an associated chain of overlapping quasi-rectangles. Assume that the constant C_{D_i} for the div$_p$ problem on $\tilde{\Omega}_i$ can be bounded by $C_{D_i} \leq C_D \frac{\delta_{M_i}}{\delta_{m_i}}$. Then, for any $1 < p < \infty$ and $f \in L_0^p(\Omega)$ there exists a solution $\mathbf{u} \in W_0^{1,p}(\Omega,\sigma^{-p})^n$ of div $\mathbf{u} = f$ such that*

$$\|D\mathbf{u}\|_{L^p(\Omega,\sigma^{-p})} \leq C\|f\|_{L^p(\Omega)}, \qquad (4.6.5)$$

where the weight $\sigma(\mathbf{x})$ depends only on the first variable and can be defined as $\sigma(x,y) := \frac{1}{\varphi'(x)}$.

Proof. We use now Theorem 4.11. Since the weight σ was treated in the previous corollary we only need to bound by below $\bar{\omega}$ defined in (4.5.12). We write $\tilde{\Omega}_{2i} = \Omega_i$ and recall properties of Definition 4.5.2 and Lemma 4.5, then

$$\bar{\omega}|_{\Omega_i} = \frac{|B_i|}{|W_i|} \geq C \frac{|R_i|}{\sum_{j \geq i} |\Omega_j|} \geq C \frac{|R_i|}{\sum_{j \leq i} |R_j|} \geq C(1 - \alpha)$$

where we used in the last inequality the decay property $|R_{j+1}| \leq \alpha |R_j|$ derived in the previous corollary. $\square$

Remark 4.15. Since a profile function is essentially constant in each Ω_i of a generalized external cusp, it is possible to "move" part of the weight σ from one side of the respective inequality to the other. Moreover, an arbitrary weight can be added to both sides if it behaves as constant in each Ω_i. We keep our presentation as clean as possible avoiding such a general treatment. The interested reader could find more about it in [4] with the warning that notation and definitions used there do not comply in general with the ones used here.

4.7 Comments and References

Many proofs of the Korn inequality have been given since Korn's original works [67, 69]. Friedrichs [43] was unable to reproduce Korn's arguments for the second case and proved by himself this inequality for *smooth* enough domains. Actually, in [43] it is first proved (3.0.1) for a class of regular functions **u** which in addition to (3.1.1) satisfy the equation $\Delta \mathbf{u} + \nabla \operatorname{div} \mathbf{u} = 0$. This is called by Friedrichs the *main case* of the inequality, since it allows to reduce the second to the simpler first case. After Friedrichs and to the middle seventies different arguments were introduced providing alternative approaches that in some cases allowed to deal with less regular domains [53, 54, 86, 49] (see the concise survey of Fichera [41]). After that, less regular functions are also incorporated, by means of appropriate weighted spaces, allowing to deal with highly irregular domains. Some of these techniques resemble aspects already mentioned in Chapter 1 in connection with Barrow's rule (1.0.5). For instance, relying on (1.0.6) it is possible to write the first derivatives of a vector field **u** as an average of derivatives of $\boldsymbol{\varepsilon}(\mathbf{u})$ in a cone (see, for instance, [64]). In this way the Korn inequality follows by using continuity properties of singular integral operators. This idea follows closely Calderón's extension method [18] (see also [5, 6]) and applies naturally for Lipschitz domains, since they enjoy the uniform cone property. In this regard, the connection between Korn's inequality and extension procedures was also exploited in a different manner by Nitsche, who using elementary arguments [81] proved (3.1.3), for Lipschitz domains and $p = 2$, modifying appropriately the extension operator due to Stein [84]. The same line of reasoning is applied for a more general class of extension domains in [38], where the authors proved, using a modification of the extension operator given by Jones [62], that (3.1.3) holds for the class of the so-called *uniform domains*. Even though this relationship between Sobolev extension and Korn inequality has proved to be useful, it failed to tackle the problem in the more general setting. Indeed, using that $Korn_p$ holds for a half of a disk it is easy to see that it also holds for the following domain

$$D_1 \setminus \{(x,y) \in \mathbb{R}^2 : 0 \leq x \leq 1, y = 0\}, \qquad (4.7.1)$$

where D_1 stands for the unitary disk. Nevertheless it is clear that in (4.7.1) does not define an extension domain, that is, it is not possible in general to extend to $\mathbb{R}^2$ functions in Sobolev spaces defined in D_1. Actually, uniform domains are strictly contained in the class of John domains. Loosely speaking, the former are characterized using "cigars" [78] joining two arbitrary point of Ω instead of a twisted cone emerging from a central $x_0 \in \Omega$.

Coming back to less irregular domains we mention [65, 66] where star-shaped domains are treated (we already cited them before as a source to derive the *Korn* inequality from improved versions of the Poincaré inequality). Another approach for Korn's inequalities on Lipschitz domains can be found in [26, 25] (actually assuming in addition that Ω is simply connected) where it is exploited the connection between the so-called Saint-Venant compatibility condition

$$\partial_{lj}e_{ik} + \partial_{ki}e_{jl} - \partial_{li}e_{jk} - \partial_{kj}e_{il} = 0 \qquad 1 \leq i,j,k,l \leq n,$$

the Poincaré's and Lions' lemmas. In particular, the Saint-Venant compatibility condition characterizes symmetric matrix fields as a linearized strain tensors of an appropriate vector field. In this way, it can be regarded as a matricial version of Poincaré's lemma which characterizes irrotational vectors fields as gradients. In turn, the latter is equivalent, under the aforementioned hypotheses on Ω, to Lion's lemma [63].

Explicit constants for Korn's inequalities are difficult to obtain even for simple domains. The survey [56] summarizes several known cases of either explicit constants or estimates for them. For instance, in [82] the constant $\frac{56}{13}$ is given for the second case in a sphere (in $\mathbb{R}^3$) and also for a spherical shell the constant can be computed in terms of inner and outer radii [8]. Actually it is known that for any domain the best constant in $\mathbb{R}^3$ is larger than 4 [57] and, to the best of our knowledge, remains as an open problem to decide whether the lowest value is attained for a ball. Similar results are given on $\mathbb{R}^2$, as an example the constant for an ellipse with semiaxes $a \leq b$ takes the form $2(1 + \frac{a^2}{b^2})$. In [56], constants for the main case and the incompressible case (the constraint $\text{div} u = 0$ is added to the second case, see also [55]) are also discussed. In [57], the constants of the second case of Korn's inequality in dimension 2 are related to those of Friedrichs inequality (2.5.4) and Babuška-Aziz, in the form $K = 2C = 2(1 + \Gamma)$ with K, C, and Γ being the constants for Korn, Babuška-Aziz, and Friedrichs, respectively. In particular upper-bounds are obtained for star-shaped domains in this case. For star-shaped domains in dimension n, the behavior of the constants is studied in [65]. Taking into account that $Korn_p$ holds on John domains it is interesting to know if the inequality could hold on a larger class of domains. An interesting recent result is given in [59] where it is proved that for a bounded domain Ω, satisfying the *separation property*, $Korn_p$ holds if and only if Ω is a John domain.

Concerning domains for which the inequality fails to hold the first counterexample for the second case of the Korn inequality and div_2 comes, to the best of the authors' knowledge, from Friedrichs himself. In [42], Friedrichs shows that, under suitable assumptions on Ω, there exists a constant C depending only on Ω, such that (2.5.4) holds. It is not difficult to see that, if Ω is simply connected, the Friedrichs inequality can be derived from both div_2 or $Korn_2$. However, in that paper, Friedrichs showed that the estimate (2.5.4) is not valid for some simply connected domains which have a quadratic external cusp (see [42, page 343]). As a consequence div_2 does not hold on such an Ω (and neither does $Korn_2$). Another counterexamples for two and three dimensional cases can be found in [48] and [87], respectively.

Besides external cusps, treated in this chapter, similar results are generalized in [58] to $s - John$ domains. In [75], using the notion of trees of domains (that generalizes the concept of chains of overlapping domains), a proof for a weighted version of div_p is given in Hölder α domains. This result is also proved in [35] following a different approach.

Appendix A
Basic Equations of Continuum Mechanics

In this appendix we give a quick review of two standard subjects in continuum mechanics: the equations of linearized elasticity and the Stokes equations. These topics have played a fundamental role in drawing the attention of the mathematical community to the subjects treated in this book. This short presentation is given for the sake of completeness and it is not rigorous. Interested readers should look at the available literature on this subject [24, 51, 23, 77, 72].

Let $\mathscr{O}$ be a continuum - such as a solid body or a fluid - identified with the closure of a bounded domain $\Omega \subset \mathbb{R}^3$ or with an indexed collection of bounded domains $\Omega_t \subset \mathbb{R}^3$. Sometimes we also write $\varphi(\overline{\Omega}, t) = \overline{\Omega}_t$ where φ, called the *motion* or *deformation mapping*, describes where each point of $\mathscr{O}$ is placed at time t.[1] The deformation field φ is assumed to be regular enough, orientation preserving, and injective in Ω in order to avoid interpenetration of matter. From the particle point of view $\varphi(\cdot, t)$ can be regarded as a trajectory and as a consequence its velocity field computed as $\mathbf{v}(\varphi(\cdot, t), t) = \frac{\partial \varphi(\cdot, t)}{\partial t}$. A straightforward application of the chain rule in the previous expression gives the well-known form of the particle acceleration as the *total or material derivative* of v, that is $D_t \mathbf{v} := \partial_t \mathbf{v} + \mathbf{v} \cdot \nabla \mathbf{v}$.

The deformation or motion of $\mathscr{O}$ due to forces exerted on it depends upon basic *balance and conservation laws*. Different kinds of external volume forces -like those given by gravitational or electromagnetic fields- may act on $\mathscr{O}$ and a generic density field $\mathbf{g} : \Omega_t \to \mathbb{R}^3$ is used here to represent any of them. Another kind of forces that should be taken into account are due to internal stresses arising between neighboring portions of $\mathscr{O}$. The description of these forces relies on a key concept that goes back to Cauchy and it is the so-called *stress tensor* $\sigma : \Omega_t \to \mathbb{R}^{3 \times 3}$. For any arbitrary portion $\mathscr{P}$ of $\mathscr{O}$ with a smooth enough boundary it is possible to assume [24, 77] that the density of forces exerted on $\mathscr{P}$ by parts of $\mathscr{O}$ surrounding $\partial \mathscr{P}$ are of the form $\tau(x) := \sigma(x)\eta$ being η the outer normal vector to $\partial \mathscr{P}$ at $x \in \partial \mathscr{P}$.[2]

---

[1] The names *reference* and *deformed configuration* for $\overline{\Omega}$ and $\overline{\Omega}_t$, respectively, are customary.

[2] The existence of a *stress vector* of the form $\tau = \tau(x, \eta)$ is a basic *assumption* in continuum mechanics. As it was early noticed by Cauchy, balance laws imply the linear dependence of τ on η

© The Author(s) 2017
G. Acosta, R.G. Durán, *Divergence Operator and Related Inequalities*,
SpringerBriefs in Mathematics, DOI 10.1007/978-1-4939-6985-2

Consider a piece of material $\mathscr{P}$ associated to a time dependent subdomain $P_t \subset \Omega_t$. Balance laws of - linear and angular- momentum state

$$\frac{d}{dt} \int_{P_t} \rho \mathbf{v} dx = \int_{P_t} \mathbf{g} dx + \int_{\partial P_t} \sigma \eta ds, \tag{A.0.1}$$

and

$$\frac{d}{dt} \int_{P_t} x \times \rho \mathbf{v} dx = \int_{P_t} x \times \mathbf{g} dx + \int_{\partial P_t} x \times \sigma \eta ds, \tag{A.0.2}$$

where $\times$ stands for the cross product and ρ represents the density of mass function associated to the continuum.

From these balance equations it follows the symmetry of σ. Indeed, consider, for instance, a situation of static equilibrium,[3] in this case $\mathbf{v} = 0$. As a consequence, the left-hand side of (A.0.1) and (A.0.2) vanish. Moreover, time dependence can be dropped in the domain of integration and therefore we can set $P = P_t$. Integration by parts -component by component- in the expression $\int_{\partial P} \sigma \eta ds$ and the arbitrariness of P lead to the fundamental relation

$$- \text{Div } \sigma = \mathbf{g}, \tag{A.0.3}$$

where the operator Div should be understood in a row-wise sense.

Writing now

$$\sigma_{ij} - \sigma_{ji} = e_j \sigma_i - e_i \sigma_j = \nabla x_j \sigma_i - \nabla x_i \sigma_j,$$

where σ_i stands for the $i-th$ row of σ, we obtain after a further integration by parts

$$\int_P \sigma_{ij} - \sigma_{ji} = - \int_P (x_j \text{Div } \sigma_i - x_i \text{Div } \sigma_j) \, dx + \int_{\partial P} (x_j \sigma_i \eta - x_i \sigma_j \eta) \, ds.$$

Then, (A.0.2) and (A.0.3) say that $\int_P \sigma_{ij} - \sigma_{ji} = 0$ and the symmetry of σ follows since P is arbitrary.

In order to go further let us recall a basic trick that takes advantage of the fact that certain physical quantities remain invariant under particular space transformations. This feature usually gives some insight about the underlying quantity.

Consider, for instance, a *linear* function $\mathbf{L} : \mathbb{R}^{3 \times 3} \to \mathbb{R}^{3 \times 3}_{sym}$, such that

$$\mathbf{L}(\mathbf{QMQ}^T) = \mathbf{QL(M)Q}^T, \tag{A.0.4}$$

[24, 77]. Notice that $\sigma \eta$ *is not normal* to $\partial \mathscr{P}$ in general. The components σ_{ij} give the projection of σe_j onto e_i and are called shear or tangential (resp. normal) stresses if $i \neq j$ (resp. $i = j$).

[3] Similar calculations can be carried out in the general, nonstatic, case.

for any *orthogonal matrix* $\mathbf{Q}$. Such an $\mathbf{L}$ is called isotropic and in particular, equation (A.0.4) implies the invariance of $\mathbf{L}$ under body rotations.[4] It is easy to see that $\mathbf{L}(\mathbf{M}) = 0$ for any $\mathbf{L}$ obeying (A.0.4) and any skew-symmetric matrix $\mathbf{M}$. Indeed, looking at the proof of Lemma 3.3, we observe that eigenvalues of $\mathbf{M}$ are of the form $\{0, i\lambda, -i\lambda\}$ with $\{0, \lambda, -\lambda\} \subset \mathbb{R}$, eigenvalues of the *Hermitian* matrix $i\mathbf{M}$. Taking the real and imaginary parts of the complex orthogonal eigenvectors of $i\mathbf{M}$ we get a matrix $\mathbf{Q}$ such that

$$\mathbf{Q}\mathbf{M}\mathbf{Q}^T = \mathbf{M}_\lambda := \begin{pmatrix} 0 & -\lambda & 0 \\ \lambda & 0 & 0 \\ 0 & 0 & 0 \end{pmatrix}, \tag{A.0.5}$$

therefore, thanks to (A.0.4) it is enough to show that $\mathbf{L}(\mathbf{M}_\lambda) = 0$.

In order to prove that, consider $\mathbf{Q}$ associated to the rotation that applies e_1 into e_2 and e_2 into $-e_1$ leaving e_3 invariant. Since for such a $\mathbf{Q}$ we have $\mathbf{Q}\mathbf{M}_\lambda \mathbf{Q}^T = \mathbf{M}_\lambda$, then (A.0.4) and the fact that $\mathbf{L}(\mathbf{M}_\lambda) \in \mathbb{R}_{sym}^{3\times3}$ readily show that $\mathbf{L}(\mathbf{M}_\lambda)$ is diagonal with $\mathbf{L}(\mathbf{M}_\lambda)_{11} = \mathbf{L}(\mathbf{M}_\lambda)_{22}$. To see that $\mathbf{L}(\mathbf{M}_\lambda)_{ii} = 0$ consider again (A.0.4) with the Q that rotates e_1 into $-e_1$ and e_3 into $-e_3$. In that case $\mathbf{Q}\mathbf{M}_\lambda \mathbf{Q}^T = -\mathbf{M}_\lambda$ while $\mathbf{Q}\mathbf{L}(\mathbf{M}_\lambda)\mathbf{Q}^T = \mathbf{L}(\mathbf{M}_\lambda)$ and therefore $\mathbf{L}(\mathbf{M}_\lambda) = 0$.

Taking into account that the kernel of $\mathbf{L}$ contains the set of skew-symmetric matrices, $\mathbf{L}(\mathbf{M})$ does not change if we replace $\mathbf{M}$ by its symmetric part. Therefore, in order to get more information about $\mathbf{L}$ it is enough to assume that $\mathbf{M} = \mathbf{M}^t$. For a diagonal matrix $\mathbf{D}$ with $\mathbf{D}_{ii} = \lambda_i$ we notice that a rotation sending exactly two canonical vectors into their opposites leaves $\mathbf{D}$ invariant, i.e. $\mathbf{Q}\mathbf{D}\mathbf{Q}^T = \mathbf{D}$. Using this and the fact that $\mathbf{L}(\mathbf{D}) \in \mathbb{R}_{sym}^{3\times3}$ it is easy to see that (A.0.4) implies $\mathbf{L}(\mathbf{D})_{ij} = 0$ for $i \neq j$. As a consequence $\mathbf{L}$ maps diagonal matrices into diagonal matrices. Consider in particular diagonal matrices of the form $\mathbf{L}(e_i \otimes e_i)$, being $(e_i \otimes e_i)_{jk} = \delta_i^j \delta_i^k$. Using a permutation matrix Q that leaves e_i invariant we see, thanks again to (A.0.4), that $\mathbf{L}(e_i \otimes e_i)_{jj} = \mathbf{L}(e_i \otimes e_i)_{kk}$ for $j \neq i \neq k$. Therefore,

$$\mathbf{L}(e_i \otimes e_i) = \alpha_i \mathbf{I} + \beta_i e_i \otimes e_i,$$

for appropriate constants α_i, β_i. Using once more (A.0.4) with a permutation that interchanges e_i with e_j we have $\alpha_i = \alpha_j$, and $\beta_i = \beta_j$ for any i, j. As a consequence, there exist constants $\alpha, \mu \in \mathbb{R}$ such that for any diagonal matrix $\mathbf{D}$, $\mathbf{D}_{ii} = \lambda_i$

$$\mathbf{L}(\mathbf{D}) = \lambda(\lambda_1 + \lambda_2 + \lambda_3)\mathbf{I} + 2\mu\mathbf{D}.$$

Since $\mathbf{L}(\mathbf{D})$ is diagonal for any diagonal matrix $\mathbf{D}$, (A.0.4) says that if $\mathbf{M} \in \mathbb{R}_{sym}^{3\times3}$ then $\mathbf{M}$ and $\mathbf{L}(\mathbf{M})$ can be simultaneously diagonalized. Therefore previous equation preserves its form for general symmetric matrices, that is

$$\mathbf{L}(\mathbf{M}) = \lambda\,tr(\mathbf{M})\mathbf{I} + 2\mu\mathbf{M},$$

---

[4] *Rigid deformations or movements* are given by linear transformations $\mathbf{T}(\mathbf{x}) = \mathbf{Q}\mathbf{x} + \mathbf{p}$ associated to *proper orthogonal matrices*, $\mathbf{Q}$: $\mathbf{Q}^T\mathbf{Q} = \mathbf{I}$ and $\det \mathbf{Q} = 1$. Nevertheless (A.0.4) holds for *any* orthogonal matrix if and only if it holds for any proper orthogonal matrix since $det(-\mathbf{Q}) = (-1)^3 \det(\mathbf{Q})$.

while for an arbitrary $\mathbf{M}$

$$\mathbf{L}(\mathbf{M}) = \lambda tr(\mathbf{M})\mathbf{I} + \mu(\mathbf{M} + \mathbf{M}^t). \tag{A.0.6}$$

As a first application of this we can now recall the form of the linear elasticity equations. For a complete treatment of the subject, we refer the reader to some classic text such as [24, 77] where a rigorous derivation is developed.

A.0.1 Linearized Elasticity

Under the action of external forces every solid body changes in shape and size. During the so-called elastic regime a deformed body tends to recover its original configuration due to the emergence of internal stresses. In this fashion, external and internal involved forces might balance each other and a static equilibrium eventually achieved. Since in static equilibrium the resultant force and moment must vanish, this situation fits in the context leading to (A.0.3) which provides three equations for the six unknowns in σ. Further constraints arise in the form of a *constitutive equation* in which stresses and strains are empirically related. In particular, for *elastic homogeneous* materials it is assumed that stresses depend only on the *deformation gradient*,[5] that is

$$\sigma = \mathbf{F}(\nabla\varphi), \tag{A.0.7}$$

where $\mathbf{F} : \mathbb{R}^{3\times3} \to \mathbb{R}^{3\times3}_{sym}$ is called a *response or constitutive* function. Thanks to (A.0.7), the number of unknowns agree with the number of equations although, without further assumptions, the question about solvability is still unanswered. In particular, the action of the external world over $\mathscr{O}$ involves not only volume forces but *surface forces* $\mathbf{g}_s : \Gamma_1 \subset \partial\Omega \to \mathbb{R}^3$ that need to be introduced in the model. As prescribed surface forces must be balanced, we may write

$$\sigma\eta = \mathbf{g}_s \text{ on } \Gamma_1, \tag{A.0.8}$$

being η the unitary outer normal vector to Γ_1.

In order to stay within a simple context let us assume that $\mathbf{F}(\mathbf{Id}) = 0$. This states that the *reference configuration* is free of stresses, since as in this case we have $\varphi(x) = x$. Furthermore, regarding small deformations as those close to the identity mapping, we might neglect nonlinear terms in the Taylor expansion of $\mathbf{F}$ around $\mathbf{Id}$ and restrict our attention to constitutive equations of the form

$$\sigma = \mathbf{L}(\nabla\mathbf{u}), \tag{A.0.9}$$

---

[5] The deformation gradient $\nabla\varphi$ is a key tool for describing strains arising in the deformation. Define, for instance, $C = \nabla\varphi^T\nabla\varphi$. Take a curve $\gamma : I \subset \mathbb{R} \to \Omega$ in the reference element, then the length element of the "deformed" curve $\varphi \circ \gamma$ can be written in the metric associated to C as $ds = \langle\nabla\varphi\gamma', \nabla\varphi\gamma'\rangle dt = \langle C\gamma', \gamma'\rangle dt$.

where $\mathbf{u}$ is the so-called *displacement field* $\mathbf{u}(x) = \varphi(x) - x$ and $\mathbf{L}$ a *linear mapping*.[6] Assuming invariance of stresses under rigid body rotations we fall in the setting of isotropic functions and therefore (A.0.9) can actually be written in terms of the *linearized strain tensor*[7] in the form known as the Hooke's law

$$\sigma = \mathbf{L}(\boldsymbol{\varepsilon}(\mathbf{u})), \tag{A.0.10}$$

which in the form (A.0.6) reads

$$\sigma(\boldsymbol{\varepsilon}(\mathbf{u})) = \lambda \operatorname{div}(\mathbf{u})\mathbf{I} + 2\mu\boldsymbol{\varepsilon}(\mathbf{u}). \tag{A.0.11}$$

The positive parameters λ and μ, called Lamé coefficients, are material dependent and should be determined experimentally. Coming back to (A.0.3) we finally obtain the equations of linear elasticity

$$- \operatorname{Div}\{\lambda \operatorname{div} \mathbf{u}\mathbf{I} + 2\mu\boldsymbol{\varepsilon}(\mathbf{u})\} = \mathbf{g} \quad \text{in} \quad \Omega. \tag{A.0.12}$$

Equation (A.0.12) must be furnished with appropriate boundary conditions. Standard traction conditions are given in the form (A.0.8) and unless $\Gamma_1 = \partial\Omega$, should be complemented with displacement conditions prescribing values of $\mathbf{u}$ on $\Gamma_2 = \partial\Omega \setminus \Gamma_1$. The *homogeneous* formulations involve (A.0.12) and the following boundary conditions

$$\{\lambda \operatorname{div} \mathbf{u}\mathbf{I} + 2\mu\boldsymbol{\varepsilon}(\mathbf{u})\}\eta = 0 \quad \text{on} \quad \Gamma_1 \tag{A.0.13}$$

$$\mathbf{u} = 0 \quad \text{on} \quad \Gamma_2. \tag{A.0.14}$$

Two standard cases are given by pure traction $\Gamma_1 = \partial\Omega$ and pure displacement $\Gamma_2 = \partial\Omega$ problems leading to classical variational formulations in Sobolev spaces.

Indeed, let us consider first the pure displacement case. Assuming that $\mathbf{g} \in L^2(\Omega)$ we can multiply each side of (A.0.12) by a test function $\mathbf{v} \in H_0^1(\Omega)^n$, and after integration by parts we find that

$$a(\mathbf{u}, \mathbf{v}) := 2\mu \int_\Omega \boldsymbol{\varepsilon}(\mathbf{u}) : \boldsymbol{\varepsilon}(\mathbf{v})\, dx + \lambda \int_\Omega \operatorname{div} \mathbf{u} \operatorname{div} \mathbf{v}\, dx = \langle \mathbf{g}, \mathbf{v}\rangle \quad \forall \mathbf{v} \in \mathbf{V}, \tag{A.0.15}$$

with $\mathbf{V} := H_0^1(\Omega)^n$. Existence and uniqueness of a solution $\mathbf{u} \in \mathbf{V}$ for (A.0.15) is granted by the Lax-Milgram theorem taking into account that the coercivity of $a(\mathbf{u}, \mathbf{v})$ on $\mathbf{V}$ follows from the first case of the Korn inequality.

---

[6] It is important to recall that the linear equations of elasticity can be rigorously derived by linearizing the general nonlinear equations [24, 77]. The informal and yet standard approach we follow here leads anyway to the right model.

[7] $\boldsymbol{\varepsilon}(\mathbf{u})$ can be obtained by linearizing the *Green-St Venant strain tensor*, $E(\varphi) = (\nabla\varphi^T\nabla\varphi - \mathbf{I})/2$ where the deformation mapping is related to the displacement through $\varphi = \mathbf{u} + \mathbf{I}$. Calling $C = \nabla\varphi^T\nabla\varphi$, we have that $C = \mathbf{I}$ if and only if φ is a rigid deformation [24]. In this sense E can be regarded as a measure of the distance from φ to the set of rigid movements. Moreover, a simple calculation shows -see Footnote 5- that E gives the *infinitesimal variation* of length due to the effect of the deformation.

Similarly, in the pure traction case, using $\mathbf{V} := H^1(\Omega)^n$ we are leading again to (A.0.15), due to the homogeneous condition (A.0.13) on $\partial\Omega$. Now in order to ensure existence and uniqueness the variational space should be restricted to

$$\mathscr{S} = \left\{ \mathbf{w} \in H^1(\Omega)^n : \quad \boldsymbol{\mu}(\mathbf{w})_\Omega = 0 \right\} \subset \mathbf{V},$$

in which the second Korn's inequality provides the coercivity of $a(\mathbf{u}, \mathbf{v})$.

A.0.2 The Stokes Equations

The Stokes equations represent a core subject in fluid dynamics. They can be easily derived within the basic framework introduced above. We need to take into account some few new physical ingredients. The first one of these is given by the assumption of mass conservation. This axiom can be written in the form known as the *continuity equation*[8]

$$\frac{\partial \rho}{\partial t} + \operatorname{div} \rho v = 0, \tag{A.0.16}$$

and thanks to this, the left-hand side of (A.0.1) can be straightforwardly computed. Indeed, by changing variables we have

$$\int_{P_t} \rho v\, dx = \int_P \rho(\varphi(x,t),t) \mathbf{v}(\varphi(x,t),t) J(x,t)\, dx,$$

where J stands for the Jacobian of the mapping $\varphi(\cdot,t) : P \to P_t$. Using (A.0.16) and the well-known formula $\frac{\partial J(x,t)}{\partial t} = J(x,t) \operatorname{div} \mathbf{v}(\varphi(x,t),t)$ from elementary calculus we get by direct calculation

$$\frac{d}{dt} \int_{P_t} \rho v\, dx = \int_{P_t} \rho \mathbf{D}_t v\, dx,$$

with $\mathbf{D}_t := \partial_t + \mathbf{v} \cdot \nabla$. Integration by parts in the surface integral of (A.0.1) and the arbitrariness of the domain of integration give

$$\rho \mathbf{D}_t \mathbf{v} = \mathbf{g} + \operatorname{div} \sigma, \tag{A.0.17}$$

---

[8] A more intuitive way to put it would be

$$\frac{d}{dt} \int_Q \rho\, dx = \int_{\partial Q} \rho v \eta\, dS$$

stating that the rate of variation of mass computed in a fixed region of the space Q equals the mass flow across the boundary of Q. Integration by parts on the right-hand side and the arbitrariness of Q give (A.0.16).

in lieu of the static version (A.0.3). The next physical consideration involves the nature of σ. Among different sources of internal stresses arising in fluids there is one associated with the notion of *pressure*. Forces due to pressure act across a surface surrounding a portion of fluid along the normal direction to that surface. In other words the stress vector $\sigma_p \eta$ is parallel to η. As a consequence σ_p can be described by means of single scalar function $p(x,t)$ in the form $\sigma_p = -p\mathbf{Id}$.[9] Having isolated the pressure, another common source of stresses arises in a *viscous flow* from diffusion processes through adjacent parts of fluid traveling at different speeds. Accordingly, it is taken

$$\sigma_v = \mathbf{F}(\nabla v),$$

as a constitutive equation, instead of (A.0.7), since stresses are caused by variations on the velocity field rather than the deformation field.

Noticing that at rest viscous stresses must vanish we may assume $\mathbf{F}(0) = 0$ and therefore, after linearizing around 0, we are led to consider constitutive linear equations of the form

$$\sigma_v = \mathbf{L}(\nabla v). \tag{A.0.18}$$

At this point, invariance of stresses under rigid body rotations says that $\mathbf{L}$ obeys (A.0.4) and therefore using (A.0.6)

$$\sigma_v = \lambda \operatorname{div} \mathbf{v}\mathbf{I} + 2\mu\boldsymbol{\varepsilon}(\mathbf{v}).$$

In the absence of any other source of internal stresses we can write $\sigma = \sigma_p + \sigma_v$ and from (A.0.17), we arrive to

$$\rho \mathbf{D}_t \mathbf{v} = -\nabla p + \lambda \nabla \operatorname{div} \mathbf{v} + 2\mu \operatorname{Div} \boldsymbol{\varepsilon}(\mathbf{v}) + \mathbf{g},$$

which can be written in the well-known form of the Navier-Stokes equations

$$\rho \mathbf{D}_t v = -\nabla p + (\lambda + \mu)\nabla \operatorname{div} \mathbf{v} + \mu \Delta \mathbf{v} + \mathbf{g}. \tag{A.0.19}$$

In the case of *incompressible flows* the velocity field $\mathbf{v}$ is solenoidal,[10] that is $\operatorname{div} \mathbf{v} = 0$ and after expanding $\mathbf{D}_t$, (A.0.19) reduces to

$$\rho \mathbf{v}_t + \rho \mathbf{v}\nabla \mathbf{v} = -\nabla p + \mu \Delta \mathbf{v} + \mathbf{g}. \tag{A.0.20}$$

In particular, in *homogeneous incompressible* flows the density ρ is constant as one can easily see from (A.0.16) and the dimensionless version of (A.0.20) takes the form

---

[9] The sign minus is arbitrary and says that stresses are directed inward for positive values of p. Fluids for which this kind of stresses are prevalent, in the sense that any other source of stress can be neglected, are called *ideal*.

[10] There are several equivalences for incompressibility. A natural definition could be $\frac{d}{dt}\int_{P_t} dV = 0$, meaning that the volume of any portion of fluid P remains constant along the flow. By changing variables it can be written as $\frac{d}{dt}\int_P J dV = 0$ which in turn gives $J = 1$ (since $J(\cdot,0) = 1$) or equivalently $\int_P \operatorname{div} \mathbf{v} dV = 0$, i.e. $\operatorname{div} \mathbf{v} = 0$.

$$\mathbf{v}_t + \mathbf{v}\nabla\mathbf{v} = -\nabla p + \frac{1}{R}\Delta\mathbf{v} + \mathbf{g}, \qquad (A.0.21)$$

where, for the sake of simplicity, the scaled variables are not renamed. The Reynold's number R plays a fundamental role in the expected behavior of the solutions of (A.0.21). For small Reynold numbers the diffusive term $\Delta\mathbf{v}$ prevails over the inertial component $\mathbf{v}\cdot\nabla\mathbf{v}$ giving rise to the following linear system

$$\mathbf{v}_t = -\nabla p + \frac{1}{R}\Delta\mathbf{v} + \mathbf{g}, \qquad (A.0.22)$$

widely known as the Stokes equations. In the case of *stationary flows* the velocity field $\mathbf{v}$ associated to the fluid does not depend on the temporal variable t, that is $\mathbf{v}_t = 0$ and we are led to the stationary Stokes equations which we write in the classical form that appears in the Introduction of this book (see (1.0.14))

$$\begin{cases} -\mu\Delta\mathbf{v} + \nabla p = \mathbf{g} \\ \quad\; \mathrm{div}\,\mathbf{v} \quad\;\; = 0, \end{cases} \qquad (A.0.23)$$

modeling *incompressible stationary viscous flows* at low Reynold's number.[11]

In spite of its simplicity, the Stokes equations have its own importance not only in applications but also in shedding light on more complex situations. The theory behind the problem of existence and uniqueness of solutions (A.0.23) relies on the existence of a right inverse of the divergence operator, as it is outlined in Chapter 1 for the case of homogeneous Dirichlet boundary conditions.

---

[11] Concerning the derivation of the associated weak formulation with the right treatment of the boundary conditions we refer the reader to the comments given below the equation (1.0.14) (see also [74]).

Appendix B
Powers of the distance to the boundary as A_p weights

Given a bounded domain Ω, in several applications (such as in Section 4.2) it is useful to know which powers of the distance to $\partial\Omega$ belong to the A_p class (2.2.30). More generally we consider the distance to a bounded closed set F that will be denoted by d_F.

Here we give a short proof of a result proved in [37, 52]. Our proof is based on known results in A_p theory and is a particular case of an argument given in [7].

The following two theorems are well known (see, for example, [31, Theorems 7.7 and Proposition 7.2]). Recall that the definition of the Hardy-Littlewood maximal function can be extended for locally finite Borel measures μ, namely,

$$M\mu(x) = \sup_{r>0} \frac{\mu(B(x,r))}{|B(x,r)|}$$

Theorem B.1. *Let μ be a locally finite Borel measure in $\mathbb{R}^n$ such that $M\mu(x) < \infty$ almost everywhere. If $0 \leq s < 1$, then $(M\mu(x))^s \in A_1$.*

Theorem B.2. *If $w_1, w_2 \in A_1$, then $w_1 w_2^{1-p} \in A_p$.*

Definition B.0.1 *For $0 < \alpha < n$, a bounded closed set $F \subset \mathbb{R}^n$ is α-regular if there exist positive constants C_1, C_2 such that*

$$C_1 r^\alpha \leq \mathscr{H}^\alpha(B(x,r) \cap F) \leq C_2 r^\alpha, \tag{B.0.1}$$

for every $x \in F$ and $0 < r \leq diam(F)$, where $\mathscr{H}^\alpha$ is the m-dimensional Hausdorff measure.

Given a domain $D \subset \mathbb{R}^n$ we define the $A_p(D)$ class taking the supremum in the definition (B.2) over cubes $Q \subset D$. When $D = \mathbb{R}^n$ we omit D from the notation.

Theorem B.3. *Let F be a closed α-regular set for some $0 < \alpha < n$ and $D := \{x \in \mathbb{R}^n : 0 < d_F(x) < diam(F)\}$. Then, for $1 \leq p < \infty$,*

$$-(n-\alpha) < \gamma < (n-\alpha)(p-1) \implies d_F^\gamma \in A_p(D) \tag{B.0.2}$$

© The Author(s) 2017
G. Acosta, R.G. Durán, *Divergence Operator and Related Inequalities*,
SpringerBriefs in Mathematics, DOI 10.1007/978-1-4939-6985-2

Proof. Consider the measure defined by $\mu(E) = \mathcal{H}^\alpha(E \cap F)$. Let us see that, for any $x \in D$,

$$M\mu(x) \sim d_F(x)^{\alpha-n}. \tag{B.0.3}$$

Clearly, modifying the constants, we can apply (B.0.1) for any r less than a fixed multiple of $diam(F)$. Given $x \in D$ let $\bar{x} \in F$ be such that $|x - \bar{x}| = d_F(x)$. If $r < d_F(x)$, then $\mu(B(x,r)) = 0$, while for $d_F(x) \le r < 2 diam(F)$ we have

$$\frac{\mu(B(x,r))}{r^n} \le \frac{\mu(B(\bar{x}, 2r))}{r^n} \sim r^{\alpha-n} \le d_F(x)^{\alpha-n}.$$

Taking $r = d_F(x)$ we have

$$d_F(x)^{\alpha-n} \sim \frac{\mu(B(\bar{x}, r))}{r^n} \le \frac{\mu(B(x, 2r))}{r^n},$$

and therefore,

$$\sup_{0 < r < 2 diam(F)} \frac{\mu(B(x,r))}{|B(x,r)|} \sim d_F(x)^{\alpha-n}.$$

Finally, for $r \ge 2 diam(F)$, $\mu(B(x,r))/|B(x,r)|$ is bounded by above by a constant depending only on F, and consequently, recalling that $0 < d_F(x) < diam(F)$ we obtain (B.0.3).

On the other hand, it is easy to see that $M\mu(x) < \infty$ for all $x \notin D$, and therefore, it follows from Theorem B.1 that, for $-(n - \alpha) < \gamma < 0$, $d_F(x)^\gamma \in A_1(D)$. Then, applying Theorem B.2 we obtain (B.0.2). $\square$

Appendix C
An Auxiliary Result on Hölder α Domains

We begin by recalling an equivalent characterization of Hölder α domains. Given α such that $0 < \alpha \leq 1$ we set $\gamma = 1/\alpha$.

Definition C.0.2 *A set $C \subset \mathbb{R}^m$ is an α-cusp, if there exist an $h > 0$ and some neighborhood of the origin $S_C \subset \mathbb{R}^{m-1}$ such that, in some orthogonal coordinate system $(x_1, \cdots, x_m) = (\mathbf{x}', x_m)$,*

$$C = \{(\mathbf{x}', x_m) \in \mathbb{R}^{m-1} \times \mathbb{R} : 0 < x_m < h, x_m^{-\gamma} \mathbf{x}' \in S_C\}$$

In some cases, we work also with an analogous definition but choosing another variable x_j in place of x_m.

Remark C.1. It can be seen that a bounded open set $\Omega \subset \mathbb{R}^m$ is a Hölder α domain if and only if for any $\mathbf{x}_0 \in \partial A$ there exists a neighborhood U of $\mathbf{x}_0$ such that $\mathbf{x} + C \subset \Omega$ for all $\mathbf{x} \in U \cap \overline{\Omega}$. For the Lipschitz case (i.e., $\alpha = 1$) this is proved, for example, in [50]. It is not difficult to see that similar arguments apply for $0 < \alpha < 1$.

For simplicity we assume, in our arguments, that Ω has diameter less than one (we can scale the original domain in order to satisfy this requirement).

The next lemma allows us to apply for Hölder α domains the ideas introduced in [17]. Since the proof is rather technical, we give all the details only for the two dimensional case. However, it is not difficult to see that the arguments can be extended to higher dimensions.

Lemma C.1. *If $\Omega \subset \mathbb{R}^n$ is a Hölder α domain then $\Omega^{k,t} \subset \mathbb{R}^{n+k}$ is also Hölder α*

Proof. We need to construct for any point of $\partial \Omega^{k,t}$ a neighborhood and an α-cusp in such a way that the translations quoted in Remark C.1 are contained in $\Omega^{k,t}$. In order to do that we will decompose $\partial \Omega^{k,t}$ in two parts: a middle one, consisting in a thin strip containing the set $(\partial \Omega, \mathbf{0}) := \{(\mathbf{x}, \mathbf{0}) : \mathbf{x} \in \partial \Omega\} \subset \partial \Omega^{k,t}$ and its complement. We divide the proof into three steps, the first two steps deal with the middle part of the boundary. In the last step we prove that the complement of the middle part is in fact smoother, showing that it is locally the graph of a Lipschitz function.

© The Author(s) 2017
G. Acosta, R.G. Durán, *Divergence Operator and Related Inequalities*,
SpringerBriefs in Mathematics, DOI 10.1007/978-1-4939-6985-2

1) Let $\mathbf{x}_0 \in \partial\Omega$. Since Ω is Hölder α there exists a neighborhood $U \subset \mathbb{R}^n$ of $\mathbf{x}_0$ and an α-cusp C such that $\mathbf{x} + C \subset \Omega$ for all $\mathbf{x} \in U \cap \Omega$. The aim of this part is to prove the following:

Claim 1: There exists an α-cusp $D \subset \mathbb{R}^{n+k}$ such that $(\mathbf{x}, \mathbf{0}) + D \subset \Omega^{k,t}$ for all $x \in U \cap \Omega$.

If

$$C = \{(x', x_n) \in \mathbb{R}^{n-1} \times \mathbb{R} : 0 < x_n < h, x_n^{-\gamma} x' \in S_C\},$$

we define

$$D = \{(x', x_n, x'') \in \mathbb{R}^{n-1} \times \mathbb{R} \times \mathbb{R}^k : 0 < x_n < \frac{h}{3}, x_n^{-\gamma}(x', x'') \in S_D\}$$

where

$$S_D = \{(x', x'') \in \mathbb{R}^{n-1} \times \mathbb{R}^k : x' \in S_C, |x''| < d_{\partial C}^t((\frac{h}{3})^\gamma x', \frac{h}{3})\}.$$

Observe that it is enough to show

$$D \subset C^{k,t}, \tag{C.0.1}$$

indeed, this inclusion gives

$$x + D \subset x + C^{k,t} = (x+C)^{k,t} \subset \Omega^{k,t}$$

as it is stated in *Claim 1*.

In order to show (C.0.1), let us consider $(x', a, x'') \in D$. Then, by definition, $0 \leq a \leq \frac{h}{3}$ and $a^{-\gamma}(x', x'') \in S_D$, so $a^{-\gamma} x' \in S_C$ and hence

$$(x', a) \in C. \tag{C.0.2}$$

On the other hand, from the definitions of D and S_D, we know that $|a^{-\gamma} x''| \leq d_{\partial C}^t((\frac{h}{3})^\gamma a^{-\gamma} x', \frac{h}{3})$, that is $|x''| \leq a^\gamma d_{\partial C}^t((\frac{h}{3})^\gamma a^{-\gamma} x', \frac{h}{3})$, and then we have to prove

$$a^\gamma d_{\partial C}^t \left(\left(\frac{h}{3}\right)^\gamma a^{-\gamma} x', \frac{h}{3} \right) \leq d_{\partial C}^t(x', a). \tag{C.0.3}$$

Observe that it is enough to consider the case $t = 1$, since $a < h < 1$. To simplify notation, we prove this inequality for $n = 2$ (as mentioned above, similar arguments apply in the general case). Then we can assume that $S_C = [-b, b]$ ($b > 0$) and x_0 is the origin. Let

$$f(x_1) = a^\gamma d_{\partial C} \left(\left(\frac{h}{3}\right)^\gamma a^{-\gamma} x_1, \frac{h}{3} \right) - d_{\partial C}(x_1, a), \qquad 0 \leq x_1 \leq ba^\gamma.$$

The function $d_{\partial C}$ is differentiable in $C - \{x_1 = 0\}$ and so f is differentiable in $\{x_1 \neq 0\}$. We have

$$f'(x_1) = \left(\frac{h}{3}\right)^\gamma \frac{\partial d_{\partial C}}{\partial x_1} \left(\left(\frac{h}{3}\right)^\gamma a^{-\gamma} x_1, \frac{h}{3} \right) - \frac{\partial d_{\partial C}}{\partial x_1}(x_1, a)$$

Now, let $(\eta, \xi) \in C$ with $0 \le \xi \le h$ (and then $0 \le \eta \le b\xi^{\gamma}$) and let $0 < \hat{\beta} < \frac{\pi}{2}$ be the acute angle between the axis x_1 and the line L passing through (η, ξ) which is orthogonal to the graph of $x_2 = b^{-\alpha}x_1^{\alpha}$ (see Figure C.1).

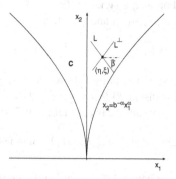

Fig. C.1 Notation used in the proof

Since $0 \le \xi \le \frac{h}{3}$, the distance $d_{\partial C}(\eta, \xi)$ is realized along the line L, and hence

$$\frac{\partial d_{\partial C}}{\partial L} = -1 \qquad \frac{\partial d_{\partial C}}{\partial L^{\perp}} = 0$$

where ∂L is understood as the outward direction along L and $L^{\perp}$ stands for the orthogonal line to L. It follows

$$\frac{\partial d_{\partial C}}{\partial x_1} = -\cos \hat{\beta}.$$

If $\hat{\beta}_a$ and $\hat{\beta}_{\frac{h}{3}}$ are the acute angles corresponding to the points (x_1, a) and $((\frac{h}{3})^{\gamma} a^{-\gamma} x_1, \frac{h}{3})$ respectively, it is easy to see that $\hat{\beta}_{\frac{h}{3}} > \hat{\beta}_a$ and then

$$f'(x_1) = -\left(\frac{h}{3}\right)^{\gamma} \cos \hat{\beta}_{\frac{h}{3}} + \cos \hat{\beta}_a \ge 0$$

since $\frac{h}{3} \le 1$. Besides $f(ba^{\gamma}) = 0$, and then $f(x_1) < 0$ for $0 \le x_1 \le ba^{\gamma}$, so inequality (C.0.3) holds for all $(x_1, a) \in C$. Hence (C.0.1) is true and the first step is proved.

2) Now we prove the following

Claim 2: Given $x_0 \in \partial\Omega$ there exists a neighborhood $V \subset \mathbf{R}^{n+k}$ of $(x_0, 0)$ and an α-cusp D such that $(x, y) + D \subset \Omega^{k,t}$ for all $(x, y) \in V \cap \Omega^{k,t}$.

Given $x_0 \in \partial\Omega$ let U and C be as in the previous step and define $V = U \times \mathbf{R}^k$. For $0 < t < 1$ we can modify, if necessary, U and C in such a way that

$$\operatorname{diam}(U) < \frac{1}{2} t^{\frac{1}{1-t}} \qquad \text{and} \qquad \operatorname{diam}(C^{k,1}) < \frac{1}{2} t^{\frac{1}{1-t}}. \tag{C.0.4}$$

By step 1 we know that $C^{k,1}$ contains an α-cusp D, hence, in order to prove *Claim 2* we will show that

$$(v,w) + C^{k,1} \subset \Omega^{k,t} \tag{C.0.5}$$

for all $(v,w) \in V \cap \Omega^{k,t}$. For such a (v,w) let $(x,y) \in (v,w) + C^{k,1}$ with $v_n - x_n \leq \frac{h}{3}$ and $y = w + \tilde{y}$. Then $|\tilde{y}| \leq d_{\partial(v+C)}(x)$, and therefore $|y| = |\tilde{y} + w|$ verifies

$$|y| \leq d^t_{\partial\Omega}(v) + d_{\partial(v+C)}(x) \leq (d_{\partial\Omega}(v) + d_{\partial(v+C)}(x))^t. \tag{C.0.6}$$

In the case $0 < t < 1$ the last inequality follows from the conditions (C.0.4). Let $\tilde{x} \in \partial\Omega$ such that $d_{\partial\Omega}(x) = \text{dist}(x, \tilde{x})$ (see Figure C.2) and let $\hat{x} = \overline{x\tilde{x}} \cap \partial C$. Finally, let $\tilde{v}$ such that $|v - \tilde{v}| = d_{\partial\Omega}(v)$ and $\overline{x\tilde{x}}$ is parallel to $\overline{v\tilde{v}}$. Then, it follows that

$$d_{\partial\Omega}(x) = |x - \tilde{x}| = |x - \hat{x}| + |\hat{x} - \tilde{x}| \geq d_{\partial(v+C)}(x) + |v - \tilde{v}| = d_{\partial(v+C)}(x) + d_{\partial\Omega}(v). \tag{C.0.7}$$

From (C.0.6) and (C.0.7) we have $|y| \leq d_{\partial\Omega}(x)^t$ and then (C.0.5) holds.

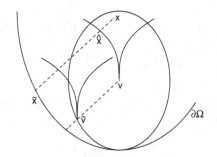

Fig. C.2 Showing that $d_{\partial\Omega}(x) \geq d_{\partial(v+C)}(x) + d_{\partial\Omega}(v)$

3) By step 2, for each $x \in \partial\Omega$ there exists a neighborhood $U_x \subset \mathbf{R}^{n+k}$ of $(x,0)$ such that $\partial\Omega^{k,t} \cap U_x$ is the graph of a Hölder α function. We have that $(\partial\Omega,0) \subset \bigcup_{x\in\partial\Omega} U_x$ and then we can extract $\{x_i\}^r_{i=1}$ such that $(\partial\Omega,0) \subset \bigcup^r_{i=1} U_{x_i}$. On the other hand, there exists $\varepsilon > 0$ such that $\{(x,y) \in \Omega^{k,t} : d_\Omega(x) < 2\varepsilon\} \subset \bigcup^r_{i=1} U_{x_i}$. In order to conclude the proof of the Lemma we will show

Claim 3: The set $\{(x,y) \in \partial\Omega^{k,t} : d_\Omega(x) > \varepsilon\}$ is locally the graph of a Lipschitz function.

Let $(x,y) \in \partial\Omega^{k,t}$ with $d_\Omega(x) > \varepsilon$. Since $y^2_1 + \ldots + y^2_k = d(x)^{2t}$ we can suppose that $|y_1|, \ldots, |y_{k-1}| < d_\Omega(x)^t$ and $y^2_k \geq \frac{d_\Omega(x)^{2t}}{k}$. Then $y^2_1 + \ldots + y^2_{k-1} = d_\Omega(x)^{2t} - y^2_k \leq (1 - \frac{1}{k})d_\Omega(x)^{2t}$. Define

$$D = \left\{ (\eta, \xi') \in \mathbf{R}^n \times \mathbf{R}^{k-1} : |x - \eta| < \frac{d_\Omega(x)}{2}, \xi^2_1 + \ldots + \xi^2_{k-1} < \left(1 - \frac{1}{2k}\right) d_\Omega(\eta)^{2t} \right\}.$$

Thus, $D \subset \mathbf{R}^n \times \mathbf{R}^{k-1}$ is a neighborhood of (x,y'), where $y = (y',y_k)$. Then we consider the function $f : D \to \mathbf{R}$ defined by

$$\xi_k = f(\eta, \xi') = \sqrt{d(\eta)^{2t} - \xi_1^2 - \ldots - \xi_{k-1}^2}.$$

It can be seen that f is a Lipschitz function. Indeed, in view of

$$d_\Omega(\eta)^{2t} - |\xi'|^2 \geq \frac{1}{2k} d(\eta)^{2t} \geq \frac{1}{2k} \varepsilon^{2t}$$

for all $(\eta, \xi') \in D$, if Λ_1 is such that $|\sqrt{a} - \sqrt{b}| \leq \Lambda_1 |a - b|$ when $a,b \geq \frac{1}{2k}\varepsilon^{2t}$, we have

$$f(\eta, \xi') - f(\alpha, \beta') = \sqrt{d_\Omega(\eta)^{2t} - |\xi'|^2} - \sqrt{d_\Omega(\alpha)^{2t} - |\beta'|^2}$$
$$\leq \Lambda_1 \left| (d_\Omega(\eta)^{2t} - |\xi'|^2) - (d_\Omega(\alpha)^{2t} - |\beta'|^2) \right|$$
$$= \Lambda_1 \left| (d_\Omega(\eta)^{2t} - d_\Omega(\alpha)^{2t}) + (|\beta'|^2 - |\xi'|^2) \right|$$

and now, if Λ_2 is such that $|a^{2t} - b^{2t}| \leq \Lambda_2 |a - b|$ for all $a,b > \frac{\varepsilon}{2}$ and $|a^2 - b^2| \leq \Lambda_2 |a-b|$ for all $a,b < \operatorname{diam} \Omega$, we get, recalling that d_Ω is Lipschitz with constant 1,

$$f(\eta, \xi') - f(\alpha, \beta') \leq \Lambda_1 \Lambda_2 \left(|d_\Omega(\eta) - d_\Omega(\alpha)| + \big||\beta'| - |\xi'|\big| \right)$$
$$\leq \Lambda \left(|\eta - \alpha| + |\beta' - \xi'| \right)$$

with $\Lambda = \Lambda_1 \Lambda_2$, as we wanted to prove. $\square$

References

1. Acosta, G., Durán, R.G., Lombardi, A.L.: Weighted Poincaré and Korn inequalities for Hölder α domains. Math. Methods Appl. Sci. **29**(4), 387–400 (2006). DOI 10.1002/mma.680. URL http://dx.doi.org/10.1002/mma.680

2. Acosta, G., Durán, R.G., López Garcia, F.: Korn inequality and divergence operator: counterexamples and optimality of weighted estimates. Proc. Am. Math. Soc. **141**(1), 217–232 (2013). DOI 10.1090/S0002-9939-2012-11408-X. URL http://dx.doi.org/10.1090/S0002-9939-2012-11408-X

3. Acosta, G., Durán, R.G., Muschietti, M.A.: Solutions of the divergence operator on John domains. Adv. Math. **206**(2), 373–401 (2006). DOI 10.1016/j.aim.2005.09.004. URL http://dx.doi.org/10.1016/j.aim.2005.09.004

4. Acosta, G., Ojea, I.: Korns inequalities for generalized external cusps. Math. Methods Appl. Sci. **39**(17), 4935–4950 (2016). DOI 10.1002/mma.3170. URL http://dx.doi.org/10.1002/mma.3170

5. Adams, R.A., Fournier, J.J.F.: Sobolev spaces, *Pure and Applied Mathematics (Amsterdam)*, vol. 140, second edn. Elsevier/Academic Press, Amsterdam (2003)

6. Agmon, S.: Lectures on elliptic boundary value problems. Prepared for publication by B. Frank Jones, Jr. with the assistance of George W. Batten, Jr. Van Nostrand Mathematical Studies, No. 2. D. Van Nostrand Co., Inc., Princeton, N.J.-Toronto-London (1965)

7. Aimar, H., Carena, M., Durán, R., Toschi, M.: Powers of distances to lower dimensional sets as Muckenhoupt weights. Acta Math. Hungar. **143**(1), 119–137 (2014). DOI 10.1007/s10474-014-0389-1. URL http://dx.doi.org/10.1007/s10474-014-0389-1

8. Andreou, E.; Dassios, G., Polyzos, D.: Korn's constant for a spherical shell. Quart. Appl. Math. **46**(3), 583–591 (1988)

© The Author(s) 2017
G. Acosta, R.G. Durán, *Divergence Operator and Related Inequalities*,
SpringerBriefs in Mathematics, DOI 10.1007/978-1-4939-6985-2

9. Arnold, D.N., Scott, L.R., Vogelius, M.: Regular inversion of the divergence operator with Dirichlet boundary conditions on a polygon. Ann. Scuola Norm. Sup. Pisa Cl. Sci. (4) **15**(2), 169–192 (1989) (1988). URL http://www.numdam.org/item?id=ASNSP_1988_4_15_2_169_0

10. Babuška, I., Aziz, A.K.: Survey lectures on the mathematical foundations of the finite element method. In: The mathematical foundations of the finite element method with applications to partial differential equations (Proc. Sympos., Univ. Maryland, Baltimore, Md., 1972), pp. 1–359. Academic Press, New York (1972). With the collaboration of G. Fix and R. B. Kellogg

11. Bernardi, C., Costabel, M., Dauge, M., Girault, V.: Continuity properties of the inf-sup constant for the divergence. SIAM J. Math. Anal. **48**(2), 1250–1271 (2016). DOI 10.1137/15M1044989. URL http://dx.doi.org/10.1137/15M1044989

12. Boas, H.P., Straube, E.J.: Integral inequalities of Hardy and Poincaré type. Proc. Amer. Math. Soc. **103**(1), 172–176 (1988). DOI 10.2307/2047547. URL http://dx.doi.org/10.2307/2047547

13. Boffi, D., Brezzi, F., Demkowicz, L.F., Durán, R.G., Falk, R.S., Fortin, M.: Mixed finite elements, compatibility conditions, and applications, *Lecture Notes in Mathematics*, vol. 1939. Springer-Verlag, Berlin; Fondazione C.I.M.E., Florence (2008). DOI 10.1007/978-3-540-78319-0. URL http://dx.doi.org/10.1007/978-3-540-78319-0. Lectures given at the C.I.M.E. Summer School held in Cetraro, June 26–July 1, 2006, Edited by Boffi and Lucia Gastaldi

14. Bogovskiĭ, M.E.: Solution of the first boundary value problem for an equation of continuity of an incompressible medium. Dokl. Akad. Nauk SSSR **248**(5), 1037–1040 (1979)

15. Brezis, H.: Functional analysis, Sobolev spaces and partial differential equations. Universitext. Springer, New York (2011)

16. Buckley, S., Koskela, P.: Sobolev-Poincaré implies John. Math. Res. Lett. **2**(5), 577–593 (1995). DOI 10.4310/MRL.1995.v2.n5.a5. URL http://dx.doi.org/10.4310/MRL.1995.v2.n5.a5

17. Buckley, S.M., Koskela, P.: New Poincaré inequalities from old. Ann. Acad. Sci. Fenn. Math. **23**(1), 251–260 (1998)

18. Calderón, A.P.: Lebesgue spaces of differentiable functions and distributions. In: Proc. Sympos. Pure Math., Vol. IV, pp. 33–49. American Mathematical Society, Providence, R.I. (1961)

19. Calderón, A.P.: Singular integrals. Bull. Amer. Math. Soc. **72**, 427–465 (1966)

20. Calderón, A.P., Zygmund, A.: On singular integrals. Amer. J. Math. **78**, 289–309 (1956)

21. Calderón, A.P., Zygmund, A.: Singular integral operators and differential equations. Amer. J. Math. **79**, 901–921 (1957)

22. Cattabriga, L.: Su un problema al contorno relativo al sistema di equazioni di Stokes. Rend. Sem. Mat. Univ. Padova **31**, 308–340 (1961)

23. Chorin, A.J., Marsden, J.E.: A mathematical introduction to fluid mechanics, *Texts in Applied Mathematics*, vol. 4, third edn. Springer-Verlag, New York (1993). DOI 10.1007/978-1-4612-0883-9. URL http://dx.doi.org/10.1007/978-1-4612-0883-9

24. Ciarlet, P.G.: Mathematical elasticity. Vol. I, *Studies in Mathematics and its Applications*, vol. 20. North-Holland Publishing Co., Amsterdam (1988). Three-dimensional elasticity

25. Ciarlet, P.G.: On Korn's inequality. Chin. Ann. Math. Ser. B **31**(5), 607–618 (2010). DOI 10.1007/s11401-010-0606-3. URL http://dx.doi.org/10.1007/s11401-010-0606-3

26. Ciarlet, P.G., Ciarlet Jr., P.: Another approach to linearized elasticity and a new proof of Korn's inequality. Math. Models Methods Appl. Sci. **15**(2), 259–271 (2005). DOI 10.1142/S0218202505000352. URL http://dx.doi.org/10.1142/S0218202505000352

27. Costabel, M., Crouzeix, M., Dauge, M., Lafranche, Y.: The inf-sup constant for the divergence on corner domains. Numer. Methods Partial Differential Equations **31**(2), 439–458 (2015). DOI 10.1002/num.21916. URL http://dx.doi.org/10.1002/num.21916

28. Costabel, M., Dauge, M.: On the inequalities of Babuška-Aziz, Friedrichs and Horgan-Payne. Arch. Ration. Mech. Anal. **217**(3), 873–898 (2015). DOI 10.1007/s00205-015-0845-2. URL http://dx.doi.org/10.1007/s00205-015-0845-2

29. Costabel, M., McIntosh, A.: On Bogovskii and regularized Poincaré integral operators for de Rham complexes on Lipschitz domains. Math. Z. **265**(2), 297–320 (2010). DOI 10.1007/s00209-009-0517-8. URL http://dx.doi.org/10.1007/s00209-009-0517-8

30. David, G., Semmes, S.: Quasiminimal surfaces of codimension 1 and John domains. Pacific J. Math. **183**(2), 213–277 (1998). DOI 10.2140/pjm.1998.183.213. URL http://dx.doi.org/10.2140/pjm.1998.183.213

31. Détraz, J.: Classes de Bergman de fonctions harmoniques. Bull. Soc. Math. France **109**(2), 259–268 (1981). URL http://www.numdam.org/item?id=BSMF_1981__109__259_0

32. Diening, L., Ružička, M., Schumacher, K.: A decomposition technique for John domains. Ann. Acad. Sci. Fenn. Math. **35**(1), 87–114 (2010). DOI 10.5186/aasfm.2010.3506. URL http://dx.doi.org/10.5186/aasfm.2010.3506

33. Drelichman, I., Durán, R.G.: Improved Poincaré inequalities with weights. J. Math. Anal. Appl. **347**(1), 286–293 (2008). DOI 10.1016/j.jmaa.2008.06.005. URL http://dx.doi.org/10.1016/j.jmaa.2008.06.005

34. Duoandikoetxea, J.: Fourier analysis, *Graduate Studies in Mathematics*, vol. 29. American Mathematical Society, Providence, RI (2001). Translated and revised from the 1995 Spanish original by David Cruz-Uribe

35. Duran, R., Muschietti, M.A., Russ, E., Tchamitchian, P.: Divergence operator and Poincaré inequalities on arbitrary bounded domains. Complex Var. Elliptic Equ. **55**(8–10), 795–816 (2010). DOI 10.1080/17476931003786659. URL http://dx.doi.org/10.1080/17476931003786659

36. Durán, R.G.: An elementary proof of the continuity from $L_0^2(\Omega)$ to $H_0^1(\Omega)^n$ of Bogovskii's right inverse of the divergence. Rev. Un. Mat. Argentina **53**(2), 59–78 (2012)

37. Durán, R.G., López García, F.: Solutions of the divergence and analysis of the Stokes equations in planar Hölder-α domains. Math. Models Methods Appl. Sci. **20**(1), 95–120 (2010). DOI 10.1142/S0218202510004167. URL http://dx.doi.org/10.1142/S0218202510004167

38. Durán, R.G., Muschietti, M.A.: The Korn inequality for Jones domains. Electron. J. Differential Equations pp. No. 127, 10 pp. (electronic) (2004)

39. Evans, L.C.: Partial differential equations, *Graduate Studies in Mathematics*, vol. 19, second edn. American Mathematical Society, Providence, RI (2010). DOI 10.1090/gsm/019. URL http://dx.doi.org/10.1090/gsm/019

40. Evans, L.C., Gariepy, R.F.: Measure theory and fine properties of functions, revised edn. Textbooks in Mathematics. CRC Press, Boca Raton, FL (2015)

41. Fichera, G.: Existence theorems in linear and semi-linear elasticity. Z. Angew. Math. Mech. **54**, T24–T36 (1974). Vorträge der Wissenschaftlichen Jahrestagung der Gesellschaft für Angewandte Mathematik und Mechanik (Munich, 1973)

42. Friedrichs, K.: On certain inequalities and characteristic value problems for analytic functions and for functions of two variables. Trans. Amer. Math. Soc. **41**(3), 321–364 (1937). DOI 10.2307/1989786. URL http://dx.doi.org/10.2307/1989786

43. Friedrichs, K.O.: On the boundary-value problems of the theory of elasticity and Korn's inequality. Ann. of Math. (2) **48**, 441–471 (1947)

44. Friesecke, G., James, R.D., Müller, S.: A theorem on geometric rigidity and the derivation of nonlinear plate theory from three-dimensional elasticity. Comm. Pure Appl. Math. **55**(11), 1461–1506 (2002). DOI 10.1002/cpa.10048. URL http://dx.doi.org/10.1002/cpa.10048

45. Gagliardo, E.: Caratterizzazioni delle tracce sulla frontiera relative ad alcune classi di funzioni in n variabili. Rend. Sem. Mat. Univ. Padova **27**, 284–305 (1957)

46. Galdi, G.P.: An introduction to the mathematical theory of the Navier-Stokes equations, second edn. Springer Monographs in Mathematics. Springer, New York (2011). DOI 10.1007/978-0-387-09620-9. URL http://dx.doi.org/10.1007/978-0-387-09620-9. Steady-state problems

47. Geißert, M., Heck, H., Hieber, M.: On the equation div $u = g$ and Bogovskii's operator in Sobolev spaces of negative order. In: Partial differential equations and functional analysis, *Oper. Theory Adv. Appl.*, vol. 168, pp. 113–121. Birkhäuser, Basel (2006). DOI 10.1007/3-7643-7601-5_7. URL http://dx.doi.org/10.1007/3-7643-7601-5_7

48. Geymonat, G., Gilardi, G.: Contre-exemples à l'inégalité de Korn et au lemme de Lions dans des domaines irréguliers. In: Équations aux dérivées partielles et applications, pp. 541–548. Gauthier-Villars, Éd. Sci. Méd. Elsevier, Paris (1998)

49. Gobert, J.: Une inégalité fondamentale de la théorie de l'élasticité. Bull. Soc. Roy. Sci. Liège **31**, 182–191 (1962)

50. Grisvard, P.: Elliptic problems in nonsmooth domains, *Monographs and Studies in Mathematics*, vol. 24. Pitman (Advanced Publishing Program), Boston, MA (1985)

51. Gurtin, M.E.: An introduction to continuum mechanics, *Mathematics in Science and Engineering*, vol. 158. Academic Press, Inc. [Harcourt Brace Jovanovich, Publishers], New York-London (1981)

52. Haroske, D.D., Piotrowska, I.: Atomic decompositions of function spaces with Muckenhoupt weights, and some relation to fractal analysis. Math. Nachr. **281**(10), 1476–1494 (2008). DOI 10.1002/mana.200510690. URL http://dx.doi.org/10.1002/mana.200510690

53. Hlaváček, I., Nečas, J.: On inequalities of Korn's type. I. Boundary-value problems for elliptic system of partial differential equations. Arch. Rational Mech. Anal. **36**, 305–311 (1970)

54. Hlaváček, I., Nečas, J.: On inequalities of Korn's type. II. Applications to linear elasticity. Arch. Rational Mech. Anal. **36**, 312–334 (1970)

55. Horgan, C.O.: On Korn's inequality for incompressible media. SIAM J. Appl. Math. **28**, 419–430 (1975)

56. Horgan, C.O.: Korn's inequalities and their applications in continuum mechanics. SIAM Rev. **37**(4), 491–511 (1995). DOI 10.1137/1037123. URL http://dx.doi.org/10.1137/1037123

57. Horgan, C.O., Payne, L.E.: On inequalities of Korn, Friedrichs and Babuška-Aziz. Arch. Rational Mech. Anal. **82**(2), 165–179 (1983). DOI 10.1007/BF00250935. URL http://dx.doi.org/10.1007/BF00250935

58. Jiang, R., Kauranen, A.: Korn inequality on irregular domains. J. Math. Anal. Appl. **423**(1), 41–59 (2015). DOI 10.1016/j.jmaa.2014.09.076. URL http://dx.doi.org/10.1016/j.jmaa.2014.09.076

59. Jiang, R., Kauranen, A.: Korn's inequality and John domains. Preprint (2015)

60. Jiang, R., Kauranen, A., Koskela, P.: Solvability of the divergence equation implies John via Poincaré inequality. Nonlinear Anal. **101**, 80–88 (2014). DOI 10.1016/j.na.2014.01.021. URL http://dx.doi.org/10.1016/j.na.2014.01.021

61. John, F.: Rotation and strain. Comm. Pure Appl. Math. **14**, 391–413 (1961)

62. Jones, P.W.: Quasiconformal mappings and extendability of functions in Sobolev spaces. Acta Math. **147**(1–2), 71–88 (1981). DOI 10.1007/BF02392869. URL http://dx.doi.org/10.1007/BF02392869

63. Kesavan, S.: On Poincaré's and J. L. Lions' lemmas. C. R. Math. Acad. Sci. Paris **340**(1), 27–30 (2005). DOI 10.1016/j.crma.2004.11.021. URL http://dx.doi.org/10.1016/j.crma.2004.11.021

64. Kikuchi, N., Oden, J.T.: Contact problems in elasticity: a study of variational inequalities and finite element methods, *SIAM Studies in Applied Mathematics*, vol. 8. Society for Industrial and Applied Mathematics (SIAM), Philadelphia, PA (1988). DOI 10.1137/1.9781611970845. URL http://dx.doi.org/10.1137/1.9781611970845

65. Kondratiev, V.A., Oleinik, O.A.: On Korn's inequalities. C. R.
    Acad. Sci. Paris Sér. I Math. **308**(16), 483–487 (1989). DOI
    10.1070/RM1989v044n06ABEH002297. URL http://dx.doi.org/
    10.1070/RM1989v044n06ABEH002297

66. Kondratiev, V.A., Oleinik, O.A.: Hardy's and Korn's type inequalities and their
    applications. Rend. Mat. Appl. (7) **10**(3), 641–666 (1990)

67. Korn, A.: Die eigenschwingungen eines elastichen korpers mit ruhender ober-
    flache. Akad. der Wissensch Munich, Math-phys. Kl, Beritche **36**(0), 351–401
    (1906)

68. Korn, A.: Solution générale du problème d'équilibre dans la théorie de
    l'élasticité, dans le cas ou les efforts sont donnés à la surface. Ann. Fac. Sci.
    Toulouse Sci. Math. Sci. Phys. (2) **10**, 165–269 (1908). URL http://www.
    numdam.org/item?id=AFST_1908_2_10__165_0

69. Korn, A.: Ubereinige ungleichungen, welche in der theorie der elastischen und
    elektrischen schwingungen eine rolle spielen. Bulletin Internationale, Cra-
    covie Akademie Umiejet, Classe de sciences mathematiques et naturelles (0),
    705–724 (1909)

70. Kufner, A., Persson, L.E.: Weighted inequalities of Hardy type. World Scien-
    tific Publishing Co., Inc., River Edge, NJ (2003). DOI 10.1142/5129. URL
    http://dx.doi.org/10.1142/5129

71. Ladyzhenskaya, O.A.: The mathematical theory of viscous incompressible
    flow. Second English edition, revised and enlarged. Translated from the Rus-
    sian by Richard A. Silverman and John Chu. Mathematics and its Applica-
    tions, Vol. 2. Gordon and Breach, Science Publishers, New York-London-Paris
    (1969)

72. Landau, L.D., Lifshits, E.M.: Teoreticheskaya fizika. Tom VII, fourth edn.
    "Nauka", Moscow (1987). Teoriya uprugosti. [Theory of elasticity], Edited by
    Lifshits, A. M. Kosevich and L. P. Pitaevskii

73. Lewis, J.L.: Uniformly fat sets. Trans. Amer. Math. Soc. **308**(1), 177–196
    (1988). DOI 10.2307/2000957. URL http://dx.doi.org/10.2307/
    2000957

74. Limache, A., Idelsohn, S., Rossi, R., Oñate, E.: The violation of objectivity
    in Laplace formulations of the Navier-Stokes equations. Internat. J. Numer.
    Methods Fluids **54**(6–8), 639–664 (2007). DOI 10.1002/fld.1480. URL
    http://dx.doi.org/10.1002/fld.1480

75. López Garcia, F.: A decomposition technique for integrable functions with
    applications to the divergence problem. J. Math. Anal. Appl. **418**(1), 79–99
    (2014). DOI 10.1016/j.jmaa.2014.03.080. URL http://dx.doi.org/
    10.1016/j.jmaa.2014.03.080

76. Magenes, E., Stampacchia, G.: I problemi al contorno per le equazioni dif-
    ferenziali di tipo ellittico. Ann. Scuola Norm. Sup. Pisa (3) **12**, 247–358 (1958)

77. Marsden, J.E., Hughes, T.J.R.: Mathematical foundations of elasticity. Dover
    Publications, Inc., New York (1994). Corrected reprint of the 1983 original

78. Martio, O.: Definitions for uniform domains. Ann. Acad. Sci. Fenn. Ser. A I
    Math. **5**(1), 197–205 (1980). DOI 10.5186/aasfm.1980.0517. URL http://
    dx.doi.org/10.5186/aasfm.1980.0517

79. Martio, O., Sarvas, J.: Injectivity theorems in plane and space. Ann. Acad. Sci. Fenn. Ser. A I Math. **4**(2), 383–401 (1979). DOI 10.5186/aasfm.1978-79.0413. URL http://dx.doi.org/10.5186/aasfm.1978-79.0413

80. Nečas, J.: Direct methods in the theory of elliptic equations. Springer Monographs in Mathematics. Springer, Heidelberg (2012). DOI 10.1007/978-3-642-10455-8. URL http://dx.doi.org/10.1007/978-3-642-10455-8. Translated from the 1967 French original by Gerard Tronel and Alois Kufner, Editorial coordination and preface by Šárka Nečasová and a contribution by Christian G. Simader

81. Nitsche, J.A.: On Korn's second inequality. RAIRO Anal. Numér. **15**(3), 237–248 (1981)

82. Payne, L.E., Weinberger, H.F.: On Korn's inequality. Arch. Rational Mech. Anal. **8**, 89–98 (1961)

83. Schumacher, K.: Solutions to the equation $\operatorname{div} u = f$ in weighted Sobolev spaces. In: Parabolic and Navier-Stokes equations. Part 2, *Banach Center Publ.*, vol. 81, pp. 433–440. Polish Acad. Sci. Inst. Math., Warsaw (2008). DOI 10.4064/bc81-0-26. URL http://dx.doi.org/10.4064/bc81-0-26

84. Stein, E.M.: Singular integrals and differentiability properties of functions. Princeton Mathematical Series, No. 30. Princeton University Press, Princeton, N.J. (1970)

85. Swanson, D., Ziemer, W.P.: Sobolev functions whose inner trace at the boundary is zero. Ark. Mat. **37**(2), 373–380 (1999). DOI 10.1007/BF02412221. URL http://dx.doi.org/10.1007/BF02412221

86. Ting, T.W.: Generalized Korn's inequalities. Tensor (N.S.) **25**, 295–302 (1972). Commemoration volumes for Prof. Dr. Akitsugu Kawaguchi's seventieth birthday, Vol. II

87. Weck, N.: Local compactness for linear elasticity in irregular domains. Math. Methods Appl. Sci. **17**(2), 107–113 (1994). DOI 10.1002/mma.1670170204. URL http://dx.doi.org/10.1002/mma.1670170204

Index

© The Author(s) 2017
G. Acosta, R.G. Durán, *Divergence Operator and Related Inequalities*,
SpringerBriefs in Mathematics, DOI 10.1007/978-1-4939-6985-2

Printed in the United States
By Bookmasters